Computational Sensor Networks

Thomas C. Henderson

Computational Sensor Networks

 Springer

Thomas C. Henderson
University of Utah
School of Computing
50 S. Central Campus Drive
Salt Lake City, UT 84112

ISBN: 978-1-4419-3501-4   e-ISBN: 978-0-387-09643-8
DOIISBN 10.1007/978-0-387-09643-8

springer.com

Dedication

To all those who participated in developing the ideas and systems presented here (especially to Felix Sawo, Kyle Luthy, Uwe Hanebeck, and Eddie Grant), to my family, and to the power of the senses!

Dedication

Preface

This book is the result of many years of effort in trying to understand sensors and sensor networks in a deep and meaningful way. It is also the work of many hands, colleagues all, including undergraduate and graduate students, and faculty and researchers from the University of Utah and other institutions. I thank them all for their contributions, discussions, and demonstrations of the ideas and technologies. I would also like to thank the reviewers which included: Edward Grant, Frans Groen, Yu H. Hu, Sitharama S. Iyengar, Gordon Lee, and Art Sanderson.

Sensors, of course, tie computing systems to the world by allowing access to the surroundings, and in this we aim to achieve what biological systems have. However, that acuity and clarity of perception, robustness, self-healing capability, fluid sensorimotor ability that we all experience daily is still far from realized in man-made artifacts. Thus, no matter what progress we record here in this monograph, the future holds even more exciting challenges and successes.

The ideas presented in this book are gathered around the insight that a sensor network can be fruitfully viewed as a computational science tool. That is, the sensor network is embedded in real world physical phenomena, and the better those can be modeled, the better the collection and analysis of data will be. Moreover, strong model-based methods allow data to be converted to information which is the foremost concern. We believe that the methodology presented here is fundamental in nature and can be usefully exploited in any sensor network.

Much remains to be done, and we have tried to point out research directions at the end of each chapter. Thus, this book should provide some guideposts to the future of sensor networks as well as an exposition of the current state-of-the-art in computational sensor networks. We look forward to participating in discovering that future!

Contents

Chapter 1

Introduction

Computational[1] sensor networks (CSN) provide a conceptual framework which offers insight into the design, analysis, development and execution of distributed sensing and actuation systems. The method depends on a set of models describing the constituent components:

- sensors,
- actuators,
- computation,
- communication, and
- physical phenomena.

Given a specific information goal, these models are exploited to explore the design space of solutions, including error and performance characterization. Sometimes special constraint functions must also be considered, e.g., temporal or energy limits, or optimal solutions are desired, and the CSN approach allows that as well. Cost benefit analysis may also be sought, and thus, techniques are needed to find regions of the design space which satisfy given criteria or boundary surfaces of interest.

CSN offers a unique vantage point as well with respect to the physical phenomena in which the system is embedded. Given a valid forward solution for the phenomenon of interest (e.g., the heat equation), it may be possible to formulate questions about the structure of the sensor network as inverse problems. For example, the heat equation gives rise to a set of nonlinear equations whose solution solves the sensor node localization problem (see Chapter 9).

The viewpoint of scientific computing may also be exploited to bring to bear:

- *simulation tools*: a CSN may be modeled and analyzed by means of standard simulation methods, but may also perform simulations as part of its real-time analysis in order to verify and validate the operational system.

---

[1]This chapter is an expanded version of work presented in [66] with K. Sikorski, K. Luthy and E. Grant.

T.C. Henderson, *Computational Sensor Networks*, DOI: 10.1007/978-0-387-09643-8_1,
© Springer Science+Business Media, LLC 2009

- *parallel and distributed system development paradigms*: a CSN typically consists of a set of communicating processors performing a distributed computation.

- *numerical methods*: a CSN typically solves systems of equations as part of its methodology.

- *software and systems engineering*: a CSN requires careful engineering in order to combine hardware and software to achieve the desired system goal.

Thus, it may be asked whether CSNs are something new as a research domain, or an amalgam of more well-established research areas. Our view is the following thesis:

> **Computational Sensor Networks** *offer a new scientific research opportunity in that systems may be developed which exploit strong models of the physical environment in which they operate in order to validate those models, as well as to probe the structure of the CSN as well. In this sense they are more self-aware than standard computational artifacts.*

A methodology is proposed here, and demonstrated by means of examples of its application. This involves statement of problem, definition of models, specification of requirements, development and deployment of embedded verification and validation methods[118], and analysis of performance.

The standpoint from which this work proceeds is that CSNs are measurement systems which are embedded in a continuous phenomenon for which they build or exploit models, and which can perform experiments to validate those models. There should be well-defined measurement goals, as well as error measures, and mechanisms (algorithms) to reduce the error to within a desired tolerance. Furthermore, nodes are generally viewed as equivalent; that is, all have the same computational, sensing, energy, and communication power, run the same algorithms, and are otherwise interchangeable; of course, the roles played by individual nodes in a specific computation may differ.

Finally, CSN science and engineering is firmly built on top of the efforts of the wireless sensor network computer architecture, embedded systems, compilers, database, and operating systems communities. However, the central CSN issues may generally be viewed as part of the application layer to the systems researchers.

1.1   Background

Sensor networks have received increasing attention over the last few years. For example, DARPA's SensIT program envisioned fields of cheap, long-lived, networked sensor devices. David Culler's work on sensor networks explores the rich design space of low-power processors, communication devices and sensors. NSF funded an STC Center for Embedded Network Systems headed by Deborah Estrin that developed algorithms for wireless and distributed sensing systems.

Some examples of issues addressed by these various projects include: power minimization [152, 166], self-configuration [15, 101], data handling [11, 72, 105], systems issues [43, 120, 167], and fault tolerance [167]. In general, higher-level exploitation of sensor networks applies standard sequential or distributed algorithms to the data. Some work in this area includes calibration [161] and habitat monitoring [107].

Sensor networks (*S-Nets*) are collections of (generally) non-mobile devices (*S-elements* or *SEL*'s) which can compute, communicate and sense the environment; oftentimes, they must be able to create local groups of devices (*S-clusters*). Our own work started in the late 1990's [62], and has mainly addressed the creation of an information layer on top of the sensor nodes. This includes distributed algorithms for leadership protocols, coördinate frame and gradient calculation, reaction-diffusion pattern formation, and level set methods to compute shortest paths through the net [19, 20, 55].

At one extreme, mobile robots can be provided with a wealth of on-board sensing, communication and computational resources [8, 146]; at the other extreme, robots with fewer on-board resources can perform their tasks in the context of a large number of stationary devices distributed throughout the task environment [62]. We have performed simulation and physical experiments using C and Matlab, as well as Berkeley motes, and the performance of robot tasks with and without the presence of an *S-Net* has been evaluated in terms of various measures. See [20, 19] for a more detailed account.

This approach can be exploited widely and across several scales of application; e.g., from robots inside buildings to robots fighting forest fires. If mobile robots are used to fight forest fires, there may be several hot spots to extinguish or control. If sensor devices can be distributed in the environment, then their values and gradients can be used to direct the behavior of fire fighting robots and to transport fire extinguishing materials from a depot to the nearest fire source. During this movement to and from the fire, collision avoidance algorithms can be employed. Sometimes coördinated activities are necessary and communication models are also important.

In our previous work, we provided models for various components of the study: (1) mobile robots with on-board sensors, (2) communication, (3) the *S-Net* (includes computation, sensing and communication), and (4) the simulation environment. We have developed algorithms in the simulation environment for the *S-Net* which perform coöperative computation and provide global information about the environment. Local and global frames are defined and created. A method for the production of global patterns using reaction-diffusion equations has been described and its relation to multi-robot cooperation demonstrated. In addition, we have shown how to compute shortest paths in the *S-Net* using level set techniques [142].

The results of our simulation experiments help us better understand the benefits and drawbacks of the *S-Net*. We have shown that for behaviors of one mobile robot going to a temperature source, and multiple mobile robots surrounding a temperature source, in the ideal situation (which means no noise), the *S-Net* approach may cost more than the non-*S-Net* system. But when noise is added in, which is more realistic, the *S-Net* system is more robust than the non-*S-Net* system. For the task of multiple mobile robots going back and forth to a temperature source, there are thresholds above which the *S-Net* system outperforms the non-*S-Net* system.

Some drawbacks of sensor networks include the need to conserve power and not run all the nodes all the time (partial data), and sensors are noisy (sometimes return the wrong value). In order to address sensor networks in a comprehensive manner, the sensor network community has initiated a research program that includes work in the areas of sensor network architectures, programming systems, reference implemen-

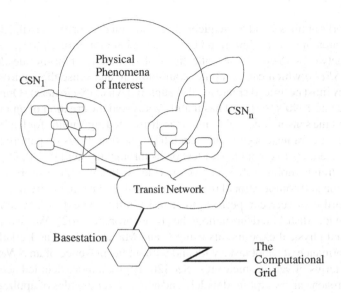

Figure 1.1: *Computational Sensor Network* Large-Scale Utilization Paradigm (adapted from [66]).

tations, hardware and software platforms, testbeds and applications. We explore the impact of a computational science approach on all these aspects of sensor networks, and show that much benefit can be derived [56, 57].

1.2   The CSN Approach

Exploiting sensor networks involves understanding algorithmic and engineering issues of real-world devices, and making both raw and processed data readily accessible to humans. In the following chapters, a general paradigm (CSN) for sensor network design and development is described, as well as a set of specific techniques for use in CSNs.

The *Computational Sensor Network* (CSN) application domain is displayed in Figure 1.1. Physical phenomena of interest are monitored by a set of CSNs, each with its own models. CSN_i produces its results (as specified by the requirements) which are passed along to other CSNs as well as to the general computational grid. These results may provide information for observers, decision makers, or may provide dynamic data for large-scale, multi-physics simulations. Figure 1.2 shows most of the system components and physical phenomena involved in a sensor network's operation. As shown in the figure, a CSN includes hardware (the *SELs*) and models may exist for power usage, fault tolerance, computational costs, etc. RF is the key issue for communication models, and making these accurate is difficult. Sensor models are essential and should be updated as time passes (e.g., bias, drift, error, etc.). Software components exist and important concerns include: correctness, numerical stability, convergence, accuracy, computational complexity, and how error and uncertainty are handled and interact from the various components. The physical phenomenon must be understood well enough

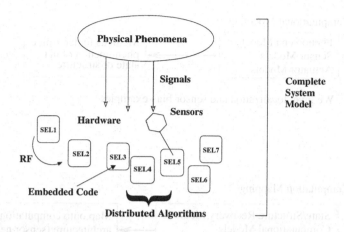

Figure 1.2: Aspects of a Computational Sensor Network.

at least to the first order, and this may involve PDE or statistical models. Finally, it is often necessary to provide some evaluation of the entire system, and this means developing models that can be used together in a correct way. This is a complicated and broad problem domain, and our goal is to provide tools to allow relevant aspects to be modeled and accounted for in developing the solution to a sensing problem.

In order to meet these analysis and system development aspects, we believe that two major issues must be addressed by the CSN system development framework (see Figure 1.3):

1. **Computational Modeling**: It is necessary to develop a framework within which it is possible to define models of physical phenomena of interest, as well as sensors and actuators, and to produce computational methods to determine state or structure of either the monitored system or the sensor network itself.

2. **Computation Mapping**: Given a method developed in (1), it is necessary to combine it with a model of the sensor network, and a set of verification and validation requirements to produce a set of executable tasks which can be mapped onto the sensor network architecture as well as a wider computational grid.

The layout of an individual CSN is shown in Figure 1.4.

CSNs provide a sensor network programming paradigm built from a combination of (1) scientific computing practice (e.g., see [87]), and (2) the Instrumented Logical Sensor methodology [31]. This combination permits the construction of qualitatively different applications by incorporation of the specific models for the phenomena being monitored, the sensors and actuators deployed, and the requirements imposed.

The rest of this book lays out the essentials of the CSN approach. Chapter 2 gives a brief detailed example of the simulation framework and describes what is meant by verification and validation. Chapter 3 gives an optimal sensor network leadership protocol. Coördinate frame development and gradient calculation algorithms are given in Chapter 4. Chapter 5 describes pattern formation using reaction-diffusion and level set applications in the CSN framework. Chapter 6 provides a complete sample simulation

(1) Computational Modeling

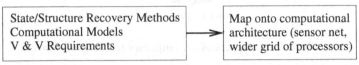

We give: localization and sensor bias examples

(2) Computation Mapping

| State/Structure Recovery Methods
Computational Models
V & V Requirements | → | Map onto computational
architecture (sensor net,
wider grid of processors) |

We give: examples from our robotics methods

Figure 1.3: *Computational Sensor Network* System Development Framework(adated from [66]).

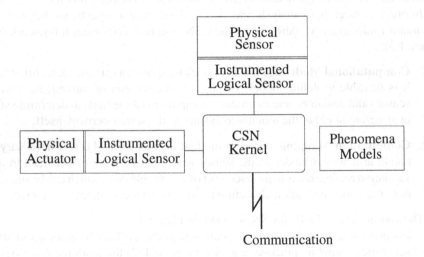

Figure 1.4: Basic *Computational Sensor Network* Layout (adated from [66]).

scenario involving mobile robots and sensor networks. Chapter 7 turns to computation mapping - that is, it provides a methodology for mapping computational models onto distributed sensor network systems while providing system support for verification and validation. Finally, Chapters 8 and 9 explain how computational models can be exploited to probe the structure of the physical phenomenon and of the sensor system as well, and in particular, the sensor node localization problem is solved.

Chapter 2

CSN: Overview of Approach

2.1 Scenario: Monitor Temperature

We start with a simple problem and cover it in detail to illustrate the ideas behind the CNS approach. We first propose a computational model for temperature variation during a 24-hour period. This model is then incorporated into a one-*SEL S-Net* in order to report any period during which the sampled temperature values are invalid with respect to (1) the temperature model, or (2) the sensor model. A detailed discussion of the simulation is given in order to facilitate the understanding of more complicated scenarios that appear in later chapters.

Problem 1: Monitor the temperature at regular intervals at a specified location with a mote, and make sure the data satisfies the local temperature model; i.e., that the measured temperature is in agreement with the temperature phenomenon and sensor models.

Problem 2: Monitor the temperature at regular intervals at a specified location with a mote, and make sure the data satisfies the sensor model; i.e., in this case that the noise is standard Gaussian.

A detailed engineering analysis requires more constraints in order to provide a real solution to this problem (including, for example, financial cost, etc.), and we have therefore assumed that the system is comprised of a standard sensor node (e.g., the Berkeley mote) which will be programmed to take temperature samples at regular intervals and transmit (i.e., wireless broadcast) a report if those values invalidate the temperature model. A solution will be developed which minimally solves the problem statement.

Before developing a physical solution, it is prudent to perform a simulation analysis. Simulation helps us determine whether the solution works as intended, helps get answers to quantitative questions, and helps make comparisons between designs. In this case, the main question to be answered by the CSN as it runs is:

T.C. Henderson, *Computational Sensor Networks*, DOI: 10.1007/978-0-387-09643-8_2,
© Springer Science+Business Media, LLC 2009

- Is the temperature model validated by the sample temperature readings?

- Is the sensor model validated by the sample temperature readings?

We next develop a simple computational model for temperature variation during a
24-hour period. This model is then exploited by an *S-Net* to test these issues.

2.2   Models

The design of a system requires models of the major components of the system. This
will also allow for a straightforward development of a simulation when desired.

2.2.1   Temperature Phenomenon Model

Table 2-1 gives a set of 24 times and temperatures recorded at the Salt Lake City, UT
airport taken between the hours of noon 17 June 2008 to 11am 18 June 2008. Using
Matlab's *polyfit* function, we determine the best cubic polynomial, T, to approximate
the data to be ($t = 1 : 24$):

$$T(t) = 0.0207t^3 - 0.7099t^2 + 5.0065 * t + 84.4348 \qquad (2.1)$$

Table 2-1. Time and Temperature (Salt Lake City, June 17-18, 2008).

Time	12	13	14	15	16	17	18	19	20	21	22	23
Temperature (F)	90	91	93	92	93	96	94	92	88	82	78	77
Time	24	01	02	03	04	05	06	07	08	09	10	11
Temperature (F)	79	73	68	68	65	63	66	66	71	74	78	79

Figure 2.1 shows this polynomial overlaid on the data. The function T characterizes
the exact temperature at every time instant.

2.2.2   Temperature Sensor Model

The temperature sensor is assumed to provide the temperature value plus some noise.
The noise value is sampled from a normal distribution with zero mean and variance one
(i.e., the standard normal distribution). This is expressed as:

$$\omega \sim \mathcal{N}(0, 1)$$

where ω is the noise. The general normal distribution with mean μ and variance σ^2 is
defined as:

$$\mathcal{N}(\mu, \sigma^2) = \frac{1}{\sigma\sqrt{2\pi}} exp^{-\frac{(x-\mu)^2}{2\sigma^2}}$$

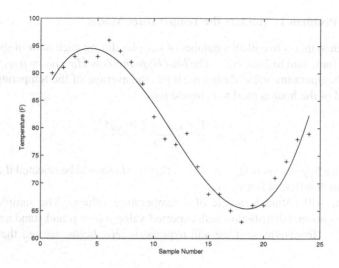

Figure 2.1: Temperature as a Function of Time.

Given this sensor model and a time during the day, sample data will be generated in the simulation by adding the temperature value produced by the model (the cubic polynomial) and a sample value drawn from the standard normal distribution:

$$T^*(t) = T(t) + \omega \qquad (2.2)$$

These two models (temperature phenomenon and sensor) can be used in the *S-Net* to monitor their validity based on the actual samples. Of course, this is a very simple model and only used to demonstrate the CSN methodology; more sophisticated temperature models (e.g., PDEs) would be needed in a realistic scenario.

2.2.3 CSN Design

In order to test the validity of the models, we will compare the sampled data to the model. If we only had Eqn (2.1), then the absolute value of the difference between the samples and the model values would provide information about the quality of the model:

$$Err(t) = \mid T^*(t) - T(t) \mid \qquad (2.3)$$

Over a period of time, samples can be accumulated and statistics of the data computed. For example, the $max\{Err\}$ might provide a good indicator of dissimilarity between the samples and the model. However, given that we have a statistical sampling process in the definition of temperature values (i.e., Eqn (2.2)), a statistical approach is therefore warranted to test the sample data.

Solution to Problem 1: Validate the Temperature Model

Our approach to this is to collect a number of samples during each hour of the day (i.e., midnight to 1am, 1am to 2am, etc.). The *Null Hypothesis* is $H_0 : \mu_0 = \mu_M$, where μ_M is the mean temperature value during the hour; the average of the temperatures at the start and end of the hour is used to compute μ_M:

$$\mu_M = \frac{T(t_{start} + T(t_{end})}{2}$$

The alternative hypothesis is $H_a : \mu \neq \mu_M$; that is, H_0 should be rejected if the sample mean $\overline{x}$ is much different from μ_M.

The mote will obtain a sample of n temperature values. The sample mean, $\overline{x}$ should have a normal distribution with expected value $\mu_{\overline{X}} = \mu$ and standard deviation $\sigma_{\overline{X}} = \sigma/\sqrt{n}$. To determine if the null hypothesis, H_0, holds, we use the following statistic:

$$z = \frac{(\overline{x} - \mu_M)}{\sigma/\sqrt{n}} \tag{2.4}$$

The rejection region for the level 0.05 test is that $z \leq -1.96$. or $z \geq 1.96$. (See [33] for a complete treatment of the statistics used here.)

Solution to Problem 2: Validate Sensor Model

In order to test the validity of the sensor model, we assume the temperature model is correct. The same data collected for the solution to Problem 1 may be used here. First the difference in each sample from the value predicted by the model is calculated:

$$err(t) = samp(t) - model(t)$$

Since the sensor error model is assumed $\mathcal{N}(0, 1)$, the Kolmogorov-Smirnov test for continuous data may be used to determine how well *err* fits the $\mathcal{N}(0, 1)$ distribution (see [129] for details on this method). The Kolmogorov-Smirnov statistic D is given by:

$$D = max\{\frac{j}{n} - F(y_{(j)}), F(y_{(j)}) - \frac{j-1}{n}\}, j = 1, \ldots, n$$

where j is an index into the n sample values. F is the density function of $\mathcal{N}(0, 1)$, and $y_{(j)}$ is the jth smallest value of the samples. We calculate the *p-value* by performing a simulation of $P_F\{D > d\}$ using the uniform distribution. If the *p-value* is low, then the hypothesis that the samples are from the $\mathcal{N}(0, 1)$ is rejected. Note that this is one important aspect of CSN; namely, that some aspect of the physical context or system features may be monitored to verify the correctness of the system during execution, as well as to probe the structure of the environment. Since the system has sensors, the models it uses may be validated during execution as well, and this leads to an adaptive system.

2.2.4   System Component Models

There are no actuators in this example, and the sensor set consists of temperature sensors. To make the simulation more realistic, we will incorporate information for the Berkeley mote concerning how much time and power it takes to acquire temperature data, to broadcast and to compute. There is some data available in the literature [138], and we will exploit it here. It is known that the radio broadcast requires 0.075 seconds, and 13.8 mA; moreover, the SenseToRfm task described in [138] is close to our problem, so we can assume that 35% of the power is spent on CPU, 6% on the sensor, and 59% on the broadcast. Thus, 1 execution (at the module level) of each requires 13.8 mA for the radio, 8.2 mA for the CPU, and 1.4 mA for the sensor. Assuming about 4×10^3 instructions and 50ns per instruction, the time for the CPU is 2×10^{-4} per cycle. Assume the time for the sensor is 0.01 sec per reading (we do not include warm-up time). Table 2-2 gives the time and energy costs.

	Time (sec)	Energy (mA)
Radio	0.075	13.8
CPU	0.0001	8.2
Sensor	0.01	1.4

Table 2-2. Time and Energy Costs for Temperature Monitoring.

Here the model will be used to determine time and power consumed, and not to characterize the statistics of the reported temperature value.

2.3   Simulation

The simulation is quite straightforward; the algorithm to monitor the sample data is first developed, and then it is instrumented to gather the information of interest (e.g., time and energy costs, statistics of interest, etc.). The nature of the system, as well as the basic statistical techniques have been described above. Relevant questions to be answered by this simulation include:

- What is the tradeoff between sample size and false positive/negative errors? energy minimization?

- What threshold values are most robust?

The simulation allows us to gain insight into these issues.

2.3.1   Method

The basic approach is described above and corresponds to *Algorithm Monitor*:

Algorithm **Monitor**:

Input: sample frequency

Output: broadcasts sensor or model invalid message

  while SEL has energy

    Get samples during 1-hour interval

    if temperature model invalid

      then Broadcast(*temperature model invalid*)

    if sensor model invalid

      then Broadcast(*sensor model invalid*)

  end

To answer the questions raised requires running a set of trials with a range of values for the number of samples and the various thresholds. These must be tried with both valid sample data as well as invalid data, and the percentage of errors determined.

2.4   Verification

A model of a simple temperature validation process has been developed, and then translated into an operational Matlab code. It is quite possible that during this process errors were made. There are several types of errors including syntactic and semantic errors. A syntactic error is some form of transliteration mistake, e.g., a variable name is misspelled, the Matlab syntax is not followed correctly, etc. Most of those may be found with conventional debugging techniques, and we do not consider that further.

Semantic mistakes on the other hand are more difficult to ferret out. For example, if a rare event is added to the wrong queue, this may lead to errors which only occasionally appear (due to the random nature of the processes, there may be some nondeterminism in the system, unless the same random numbers are used repeatedly for debugging purposes).

For a general introduction to verification and validation in discrete event simulation, see [98], and for a strong view from the computational science community, see [118]. Here we will simply outline the verification process used for this simulation.

There are several important aspects to verification. First, we must ensure as well as possible that the input streams of samples of random variables are correct (i.e., are a sample from the desired random variable). It is also useful to include checks to see if known important conditions are ever violated. Another thing to check is that corner cases are handled correctly. Finally, it is good to run the code on samples where the solution is known to check that the correct solution is found.

2.4.1   Input Streams

The only input stream for this simulation is drawn from the standard Gaussian distribution for the temperature noise. Figure 2.2 gives a histogram of the noise samples from a run of the code. No χ^2 statistic is computed, but the histogram looks Gaussian. If this did not look right, then a more in-depth analysis would be called for, and appropriate

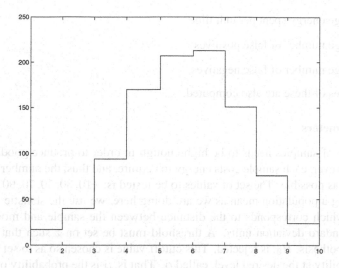

Figure 2.2: Histogram of Temperature Sample Noise.

changes in the simulation would need to be made; for example, it might be necessary to use more samples.

2.4.2 Known Result Comparison

An easy way to have a known result is to eliminate the random nature of the simulation. For example, if no noise is added to the temperature sample, then there should be no broadcast messages. If, on the other hand, the temperature is set to always return a 0 value, then every sample set should cause a broadcast.

Another type of check consists of invariants; e.g., the sample times should all fall within a single wall-clock hour; this can be checked as the code runs. The number of broadcasts reporting an invalid model should never exceed the number of sample sets; this is also true for broadcasts reporting an invalid sensor model.

2.4.3 Data, Analysis and Interpretation

Data

The simulation is run in order to gain insight concerning the number of samples required, the energy costs due to that, the threshold values of validity checking and the relationship of these to the robustness of the process. The parameters to be studied are:

- number of samples
- threshold for the temperature model validity z-statistic
- threshold for the sensor model validity p-value statistic

The statistics to be ascertained include:

- average energy spent per unit time

- average number of false positives

- average number of false negatives

The variances of these are also computed.

Input Parameters

The number of samples needs to be high enough in order to produce good statistical values. However, each sample costs energy to acquire, and thus, the number should be kept as low as possible. The set of values to be tested is: $\{10, 30, 50, 70, 90\}$.

In testing a population mean as we are doing here, we use the statistic z given in Eqn (2.4) which corresponds to the distance between the sample and model means given in standard deviation units. A threshold must be set on z such that if $z \geq c$, then the hypothesis, H_0, is rejected. This cutoff value is chosen so as to set the Type I error probability at the desired level, called α. That is, α is the probability of rejecting H_0 when it is true. Thus, the smaller α (which implies a higher threshold), the less likely a Type I error occurs. Here the set of α's is $\{0.05, 0.20\}$ corresponding to cutoff thresholds of $\{1.96, 1.28\}$, respectively.

Finally, in order to test the goodness of fit of the sample data to a $\mathcal{N}(0, 1)$ distribution using the Kolmogorov-Smirnov test statistic D, it is necessary to determine how likely the D value is, given that the samples come from $\mathcal{N}(0, 1)$ (given that H_0 is true). To that end, the *p-value* is defined as the probability of getting values of D larger than the specific d found for the sample set. This probability is estimated using simulation (and can be done off-line). In order to test the sensitivity of the *p-value*, the range for testing is $\{0.01, 0.05\}$. Thus, there are $5 \times 2 \times 2 = 20$ test cases to run. The output statistics for these are given in Table 2-1.

This data is also shown in Figures 2.3 and 2.4.

The simulation was also run with a different sensor model for the acquisition of temperatures (noise samples were taken from $\mathcal{N}(2, 3)$). Table 2-2 gives the results for this.

This data is plotted in Figures 2.5 and 2.6.

Analysis and Interpretation

As can be seen very clearly, the 10-sample version with $c = 1.96, p = 0.01$ significantly outperforms the 90-sample version since it runs around 1,030 cycles versus 115, and it makes a lower percentage of temperature model errors (1.45% vs. 1.48%) and sensor model errors are comparable (1.27% vs. 1.25%). The order of magnitude greater running time is the most significant feature.

The question arises as to why fewer samples should outperform more samples. One possibility is that since μ_M is the average between the two hourly values, then when there are fewer samples, they are closer to that value (samples are evenly spaced about the midpoint). This conjecture has not been verified.

When considering which parameters fare better for the case of a bad sensor model (i.e., the actual mean of the noise is 2, and the variance is 3), the 10-sample version

Table 2-1. Data from Simulation of *Algorithm Monitor* (*n*: number of sample; *c*: cutoff; *p*: *p*-value).

n	*c*	*p*	invalid model mean	invalid model variance	invalid sensor mean	invalid sensor variance	cycles
10	1.28	0.05	88.50	83.17	52.40	46.27	1020.90
10	1.96	0.01	89.50	98.28	10.00	7.11	1026.90
10	1.28	0.05	14.30	14.46	52.60	59.38	1031.60
10	1.96	0.01	15.00	22.89	13.20	20.62	1037.10
30	1.28	0.05	32.70	30.68	18.30	17.34	344.30
30	1.96	0.01	29.90	12.99	3.00	2.00	345.10
30	1.28	0.05	3.40	4.04	17.00	16.67	345.80
30	1.96	0.01	3.50	3.39	3.60	1.38	346.10
50	1.28	0.05	23.70	19.57	9.80	13.96	207.00
50	1.96	0.01	23.60	8.04	1.90	4.54	207.00
50	1.28	0.05	2.30	1.34	9.20	7.96	207.50
50	1.96	0.01	2.00	3.33	2.0	1.56	208.00
70	1.28	0.05	20.20	21.96	7.30	5.34	148.00
70	1.96	0.01	23.20	5.07	7.30	5.34	148.00
70	1.28	0.05	3.60	3.16	9.10	6.32	148.00
70	1.96	0.01	2.70	2.68	1.10	1.21	148.00
90	1.28	0.05	20.40	43.82	6.40	10.93	115.00
90	1.96	0.01	21.40	7.60	0.50	0.50	115.00
90	1.28	0.05	3.40	3.60	5.80	3.29	115.00
90	1.96	0.01	1.70	3.24	1.40	2.27	115.00

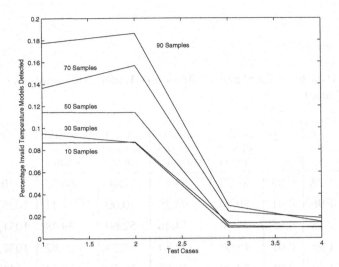

Figure 2.3: Percentage of Correctly Detected Invalid Temperature Models.

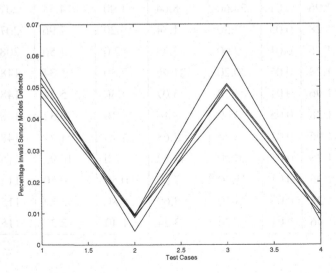

Figure 2.4: Percentage of Correctly Detected Invalid Sensor Models.

Table 2-2. Data from Simulation of *Algorithm Monitor* with Bad Sensor.

n	c	p	invalid model mean	invalid model variance	invalid sensor mean	invalid sensor variance	cycles
10	1.28	0.05	803.80	3.96	780.30	28.46	813.30
10	1.96	0.01	813.70	13.57	694.40	37.82	824.40
10	1.28	0.05	755.10	28.99	786.70	13.34	819.50
10	1.96	0.01	761.80	40.62	694.90	101.88	831.60
30	1.28	0.05	33.70	0.68	329.00	0.00	329.00
30	1.96	0.01	32.20	0.62	328.60	0.49	329.00
30	1.28	0.05	29.70	0.68	329.50	0.28	329.50
30	1.96	0.01	28.80	0.40	329.90	0.10	329.90
50	1.28	0.05	36.20	0.62	201.00	0.00	201.00
50	1.96	0.01	35.30	1.34	201.00	0.00	201.00
50	1.28	0.05	32.00	0.44	201.00	0.00	201.00
50	1.96	0.01	31.30	0.46	201.00	0.00	201.00
70	1.28	0.05	37.60	0.27	145.00	0.00	145.00
70	1.96	0.01	37.80	0.40	145.00	0.00	145.00
70	1.28	0.05	32.90	0.54	145.00	0.00	145.00
70	1.96	0.01	33.20	0.62	145.00	0.00	145.00
90	1.28	0.05	38.90	0.54	113.00	0.00	113.00
90	1.96	0.01	38.60	0.71	113.00	0.00	113.00
90	1.28	0.05	34.60	0.49	113.00	0.00	113.00
90	1.96	0.01	34.00	0.22	113.00	0.00	113.00

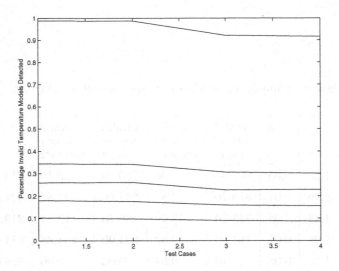

Figure 2.5: Percentage of Invalid Temperature Models with Bad Sensor Model.

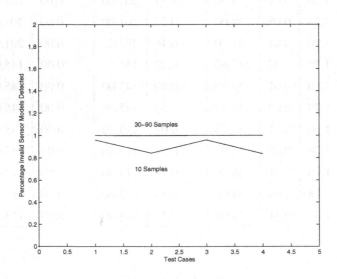

Figure 2.6: Percentage of Invalid Sensor Models with Bad Sensor Model.

with $c = 1.96$, and $p = 0.01$ still looks to perform the best in the sense that it detects an invalid temperature model 92% of the time, and invalid sensor model 84% of the time. Depending on the application, it may be warranted to use more samples, as in all the other cases, the invalid sensor model is detected 100% of the time, however, an invalid temperature model is reported only about 8.73% of the time. This may be considered better since the two different statistical tests can differentiate between a bad temperature model and a bad sensor model.

2.5 Validation

In order to validate the simulation, it would be necessary to collect some data from physical experiments and see that the models match, and that the predicted values match. In this instance, we have based the input models on experimental data from the literature and the airport temperature recording site, and checked that the simulation values reflect the actual data. As for the validation computation, it is a bit too simplistic to be accurate, but we could indeed run some physical experiments to see if the validation results for the system are accurate. If they are not correct, it would be necessary to determine why that is the case, and then modify the model to account for those things.

2.6 Summary

In this chapter, a simple simulation was developed, and the basic structure of the simulation was described. Case handling was explained, as well as how to make use of input distributions. Simulation verification and validation specifics were given, and an example of how knowledge of the physical phenomenon can be used to determine when the system is operating outside the proscribed range of operation.

Chapter 3
Leadership Algorithms

It is sometimes important to have a local leader[1] for a set of sensor nodes. Such a leader may be used as the origin of a coordinate system, as the node responsible for communication, etc. Thus it is important to have a reliable and correct method to assign nodes as leaders. In order to proceed, it is necessary to give as careful a definition of the problem – and its solution! – as possible.

The Leadership Problem: Each *SEL* has a unique integer ID (UID) and a fixed geographic location; *SELs* have a restricted broadcast range which defines a connectivity graph. The *SELs* are to be grouped into subgraphs, called *S-clusters*, such that each *S-cluster* has a leader, and the leader of each *S-cluster* has the lowest ID of all members of the *S-cluster*.

In this chapter, we describe an algorithm to solve the *S-cluster* leadership problem [55]. For a good introduction to distributed algorithms, including solutions to variations of the leadership problem and correctness proofs, see [104]. For a leadership election protocol in the context of target tracking, see [168].

The algorithm presented here is optimal with respect to the number of broadcasts, and has some very nice properties as determined on nodes whose locations are random samples from a uniform distribution in a square area. Given a set of *SELs* which have determined their neighbors:

- Each *SEL* broadcasts exactly one message during execution of the leadership protocol.

- The number of leaders is bounded by the maximum number of circles (whose radius is the broadcast range) which can be packed into a square area.

---

[1]This chapter is an expanded version of work presented in [55, 58], as well as work with Jong-Chul Park, Nate Smith and Richard Wright [64].

3.1   Leadership Protocol

We gave an algorithm to solve the *S-cluster* leadership problem [55]. For a good introduction to distributed algorithms, including solutions to variations of the leadership problem and correctness proofs, see [104]. For a leadership election protocol in the context of target tracking, see [168]. Others have introduced leadership protocols (also called cluster formation algorithms); e.g., Chan and Perrig [17] described the ACE algorithm which is an emergent algorithm to form highly uniform clusters, and Shin et al. gave a variation of that [144]. However, both of these algorithms are much more restrictive than SNL in that they require that clusters be disjoint, and thus their methods require an iterative broadcast procedure which consumes much more energy than SNL which requires only one broadcast per node to determine the leaders. The leadership problem may be defined as follows:

An *S-Net* system will be represented as an undirected graph where each node is a *SEL*. Note that the assumption is that the graph is undirected; however, this is something that must be established by a lower level algorithm (e.g, as part of the communication protocols). It is not the case, in general, that pairs of *SEL*s can receive broadcasts from one another. Each node is a distinct process and each is placed in the environment as a distinct hardware device.

Formal definitions can be given for the nodes, and this involves defining states, including start states, message generating functions, and state transitions. However, only an informal description is given here. Such a description will include *broadcast()* and *receive()* primitive functions with their associated messages. A *broadcast* sends a message to all *SEL*s within range. Proof methods typically involve either invariant assertions and a demonstration that they hold; simulations are used to explore the average case behavior.

A simple example of a leadership algorithm is the LCR algorithm which provides a basic solution to the leadership problem in a synchronous ring network [104]; it involves each process sending its UID in one direction around the ring to its neighbor; when a process receives a UID, it will throw it away if it is less than its own, resend it to its neighbor if it is larger than its own, and declare itself the leader if it is equal to its own. Our solution is related to this idea, although not the same.

The *S-Net* leadership basic algorithm (**SNL**) is executed by each node, and is as follows:

Algorithm **SNL**:

   Step 1. Broadcast own ID for a fixed time, T1.

   Step 2. Receive from other nodes, create neighbors list for a fixed time, T1

   Step 3. Create remaining nodes list (initially, neighbors)
       while not done
           if node's own ID is lower than min ID in remaining nodes list,
               then node is leader
                   broadcast cluster (self and neighbors)

       done

    else receive broadcast cluster list

      if in list

        node is not a leader

        re-broadcast list

        done

      else remove list from remaining

Note that we assume that enough time is given to Steps 1 and 2 so that each node can complete the step correctly. This will most likely be implemented as a fixed time delay in an embedded system. Also, we assume that there are communications protocols that are reliable enough to transmit the messages without loss of information, and to ensure that communication between nodes is bi-directional.

3.2 Correctness

We outline an informal argument for the correctness of algorithm **SNL**. Let $\mathcal{U} = \{1, 2, \ldots, uid_{max}\}$. The message alphabet $\mathcal{M}$ is the power set of $\mathcal{U}$, i.e., $\mathcal{P}(\mathcal{U})$.
The state of each node includes:

- my_UID_i: node i's unique UID (e.g., $my\_UID_i = i$)

- *broadcast*: a message in $\mathcal{M}$ or *null*, initially *null*

- *leader*: a Boolean, indicating whether the node is a leader, initially *false*

- *resolved*: a Boolean, indicating whether the node has resolved as either a leader or not, initially *false*

Data structures used include:

- *neighbors*: list of *SEL* neighbors, initially *null*

- *remaining*: list of *SEL* neighbors still unresolved, initially *null*

The start state for each node i is that initial set of values indicated above. For each node, the following messages are possible:

- *self*: consists of my_UID_i

- *cluster*: list of UID's that form a cluster; i.e., a leader and its neighbors

The transition function for **SNL** is defined as:

```
% Step 1 of SNL
while (timer1 > 0)
  broadcast self;
endwhile

% Step 2 of SNL   (runs concurrently with Step 1)
while (timer1 > 0)
  add_to_neighbors(receive())
endwhile

remaining = neighbors;
% Step 3 of SNL
while (not resolved)
  % Step 3.1
  if (my_UID(i) < min(remaining))
    leader = true;
    resolved = true;
    broadcast(my_UID(i), neighbors);
    exit;
  endif

  list = receive();
  % Step 3.2
  if (my_UID(i) in list)
    leader = false;
    resolved = true;
    broadcast(list);
    exit;
  endif
  remaining = remaining - list;
endwhile
```

Note that the broadcast in (3.2) has to take place so that a node i not in the cluster, but neighboring a node j in the cluster, can know that node j is resolved; this is necessary since the leader will not reach the non-cluster nodes that neighbor cluster nodes (i.e., the broadcast from the leader node will not reach node i).

The algorithm is supposed to achieve:

- (i) $leader = true$

  for any node that has the lowest UID of it and its unresolved neighbors.

- (ii) $leader = false$

  for any node that neighbors a leader.

- (iii) $resolved = true$

  for every node.

Case (i)

Suppose that node i has the lowest UID of it and any of its neighbors. Then when it finishes Step (2),

$$remaining = (nei_{i_1}\_UID, \ldots, nei_{i_k}\_UID)$$

Thus, in Step (3),

$$\forall j \quad my\_UID_i < nei_{i_j}\_UID$$

Node i then asserts itself as a leader.

Case (ii)

Suppose node i has a neighbor which eventually asserts itself a leader, say $nei_{i_m}\_UID$. Then,

$$remaining = (nei_{i_1}\_UID, \ldots, nei_{i_m}\_UID, \ldots)$$

and (3.1) is always *false* as long as node i does not assert itself as a leader. This is true because $nei_{i_m}\_UID$ will not be removed from *remaining* unless a *SEL* is declared with node i_m as a member. Eventually, node i_m will assert itself as a leader, and will broadcast a list with node i as a member. Thus, (3.2) will be true, and node i will declare itself not a leader.

Case (iii)

Every node is a leader or neighbors a leader. Thus, eventually one of cases (i) or (ii) will occur, and in each case, node i is resolved.

The algorithm is optimal with respect to the number of broadcasts, and has some very nice properties as determined on nodes whose locations are random samples from a uniform distribution in a square area. Given a set of *SELs* which have determined their neighbors:

- Each *SEL* broadcasts exactly one message during execution of the leadership protocol.

- The number of leaders is bounded by the maximum number of circles (whose radius is the broadcast range) which can be packed into a square area.

Note that Perrig's ACE algorithm requires 7 broadcasts per node per iteration of the algorithm (there may be several), while the Node Degree algorithm requires on average 1.1 broadcasts per node for each of several iterations. SNL requires only a single broadcast per node for the entire execution (after neighbors are established).

3.3 SNL Simulation

Figure 3.1 shows the result of running a simulated version of the SNL protocol on 81 *SELs* which are arranged in a 9x9 grid layout. The broadcast range for each *SEL* is circular with radius 1.1 units; this means each *SEL* can reach its 4-neighbors (distance 1), but not its diagonal neighbors (distance $\sqrt{2}$). This can be verified in the figure as each leader is a circle and *SEL* n, where n is odd, is a leader.

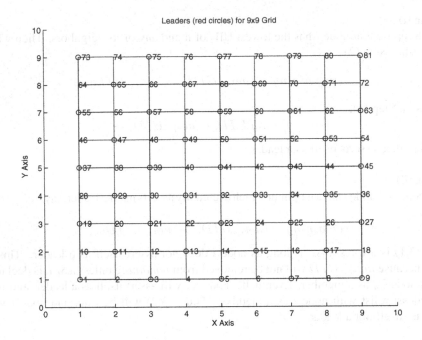

Figure 3.1: SNL Protocol Result on a 9x9 Grid with Broadcast Range 1.1 Units (adapted from [58]).

To better understand the way SNL works, consider the 4-node layout in Figure 3.2. The node locations, IDs and neighbors are given in Table 3.1. The broadcast range is 1.5 units.

Table 3.1 A Simple *SEL* Set.

Node ID	x	y	Neighbors
1	5	4	2,3
2	4	5	1,3,4
3	6	5	1,2,4
4	5	6	2,3

The nodes proceed asynchronously and at the first iteration of *Step 3*, the following occurs:

Node 1: has a lower ID than its neighbors, and will assert itself as a *leader*.

Node 2: has Node 1 as a neighbor and therefore performs a receive.

Node 3: has Node 1 as a neighbor and therefore performs a receive.

Node 4: has Nodes 2 and 3 as neighbors and therefore performs a receive.

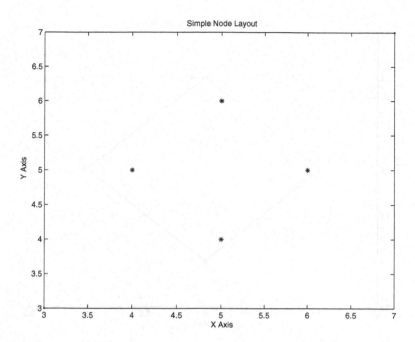

Figure 3.2: Simple *SEL* Layout to Demonstrate SNL Protocol (adapted from [58]).

Eventually Node 1 will broadcast its cluster: $[1, 2, 3]$. The other nodes will loop waiting to receive a broadcast. Nodes 2 and 3 will receive Node 1's broadcast, but Node 4 is out of Node 1's broadcast range and will not receive it.

After Node 1 broadcasts its cluster, it exits and goes to other tasks. Suppose Node 3 receives the broadcast first (this is nondeterministic); then Node 3 finds its ID in the list and asserts itself as a *follower*, re-broadcasts the list, and exits. Node 2 will eventually receive the list and assert itself as a *follower*, re-broadcast the list and exit. Eventually, Node 4 will receive the broadcast from Node 2 or Node 3. Node 4 does not find itself in the cluster $[1, 2, 3]$, and it re-assigns its *remaining* list as $[2, 3] - [1, 2, 3]$ which is the empty list. At this point, Node 4's ID is lower than anything on the list, and so Node 4 asserts itself as a *leader* and exits. Figure 3.3 shows the resulting leadership structure (Nodes 1 and 4 are *leaders* and Nodes 2 and 3 are *followers*).

3.3.1   The Simulation Logic

The SNL protocol simulation builds on the monitor simulation in the previous chapter. It is organized as follows:

Simulation Protocol:

*SEL*s are initialized as described.

Broadcast ID events are scheduled for nodes.

Receive events are scheduled for nodes.

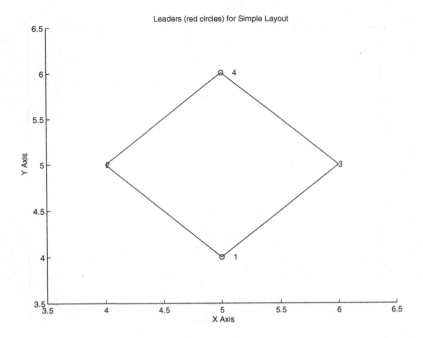

Figure 3.3: Result of SNL Protocol on Simple *SEL* Layout (adapted from [58]).

while event-queue $\neq \emptyset$ and $\exists$ unresolved nodes

    Select next event.

    Handle next event.

end.

The events are:

Broadcast ID: Broadcast ID and schedule next broadcast ID if still in Phase I (Steps 1 and 2).

Broadcast receive: Receive a broadcast and schedule next receive event if still in Phase I.

Broadcast neighbors: Broadcast neighbors list.

Broadcast cluster: Broadcast cluster list.

Receive ID: Receive ID and schedule next receive ID event if still in Phase I.

Receive cluster: Handle part of Step 3 when node is not a leader; i.e., receives cluster list and either resolves as follower if in list or otherwise subtracts received list from *remaining* and schedules new receive cluster event.

Phase I timer end: Initializes *SEL*'s *neighbors* and *remaining* lists and schedules a first execution of Step 3.1 (i.e., if leader, broadcast cluster; otherwise, schedule a receive list event).

Determine Role: Execute Step 3.1 of SNL algorithm. If *SEL* is not a leader, schedule a receive cluster event.

3.3.2 Verification

The algorithm assumes that all neighbor relations are bi-directional. A check is put into the code for this prior to starting Step 3.

Other verification checks include (1) no leader neighbors another leader, (2) every follower neighbors at least one leader, and (3) every *SEL* is resolved (i.e., is either a leader or follower).

Alternatively, this can be formulated as (1) every *SEL* is either a leader or a follower and in a cluster, (2) every follower neighbors at least one leader, and (3) every neighbor of a leader is in its cluster. This is the check performed in the code and it has been run on thousands of randomly generated networks, and correctness tested.

3.3.3 Validation

There are many sensor networks whose structure can be exploited to test validity. For example, all odd-sided unit grids numbered by row whose *SELs* have broadcast range 1.1, should have all odd nodes as leaders. Regular polygon nets with *SELs* on the unit circle and broadcast range $1.1\sqrt{2(1 - cos(\theta))}$, where θ is the angle between two adjacent points, should only have the two nearest polygon points as neighbors. Tests have been run for up to 200 without error to test validity on such polygon nets (a ring network).

3.3.4 SNL Protocol Statistics

The SNL protocol results in a structure of *leaders* and *followers*, and some of the properties of this structure are of interest. Given a set of n node locations sampled from a uniform 2-D distribution, and with randomly assigned *SEL* ID's, we study the following statistics:

- average number of leaders, and

- their spatial distribution.

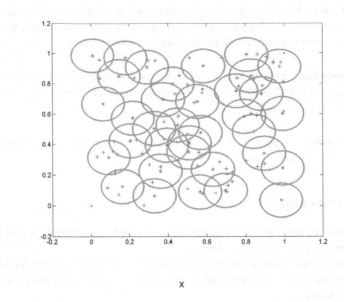

Figure 3.4: Results of SNL Protocol on 100-*SEL* Configuration.

To obtain these statistics, a suitable framework must be established. We consider *SELs* distributed randomly in the unit square, and each having the same broadcast range, r, $0 < r <= 1$. Thus, the leadership protocol structure is a function of the spatial distribution and density, and the broadcast range. Figure 3.4 shows the result of running the SNL protocol on a 100-*SEL* configuration. Figures 3.5 and 3.6 show for various values of r (1, 0.707, 0.5, 0.25, 0.1, 0.05 and 0.01) the average number of clusters per number of *SELs* (10 to 100).

The simulation protocol for a given number, n, of *SELs* and broadcast radius r, is as follows: (1) a trial consists of the generation of 200 random layouts for the *SELs* and the execution of the SNL protocol for each layout; the mean number of leaders is then computed for these 200 results; (2) 20 trials are run and the mean and variance computed for the 20 trials. As a verification check that the data is correct, the average node degree is calculated and shown to grow linearly with the number of *SELs*. No error bars are shown for the average number of leaders since the 95% confidence interval is about 0.001; thus, confidence is high for a narrow spread about the mean.

As can be seen in Figures 3.5 and 3.6, the number of leaders (and therefore clusters) approaches a limiting value for the larger radii, but continues to grow through 100 *SELs* for the smaller radii. Some interesting questions are: (1) What is the maximum number of leaders possible? and (2) Does the average approach the maximum as the number of *SELs* goes to infinity?

The first question can be posed as a circle packing problem (see [149, 153] for a good introduction to circle packing). The best solutions for packing up to 200 circles into the unit square are given in Table 13.1 in [153]; we give a selected subset in Table 3.2 here.

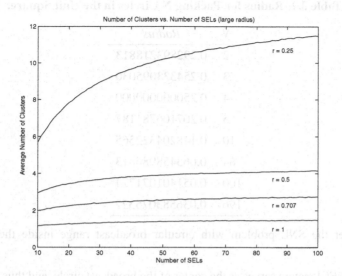

Figure 3.5: Average Number of Clusters vs. Number of *SEL*s in Network (adapted from [58]).

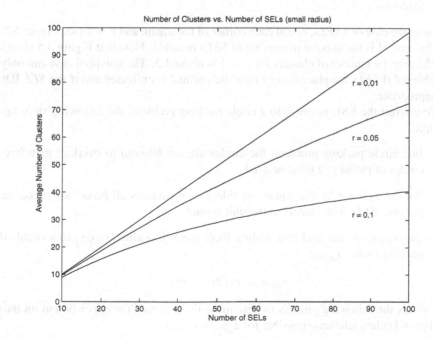

Figure 3.6: Average Number of Clusters vs. Number of *SEL*s in Network (adapted from [58]).

Table 3.2. Radius for Packing N Circles in the Unit Square.

N	Radius
2	0.292893218813
3	0.254333095030
4	0.250000000000
5	0.207106781187
10	0.148204322565
64	0.063458986813
100	0.051401071774
196	0.036583075322

Consider the SNL problem with circular broadcast range inside the circle of radius r:

1. The *SEL* location serves as the center of the broadcast circle, and thus all centers of the circles must be in the unit square. However, part of the circle may extend beyond the square.

2. No two leaders may directly communicate, and the minimum distance between leaders is r.

Consider the case of 4 *SELs*, one at each corner of the square and $r = 1$ (see Figure 3.7. For this case, 4 is the maximum number of *SELs* possible. Note that Figure 3.5 shows that the average number of clusters for $r = 1$ is about 1.5. The maximal case can only be achieved if *SELs* are placed on or near the optimal coordinates and if the *SEL* IDs are appropriate.

To convert the SNL problem to a circle packing problem, the following steps are required:

1. In a circle packing problem, the circles are not allowed to overlap; therefore, circles of radius $r/2$ must be used.

2. For the radius $r/2$, the square of side $1 + r$ contains all broadcast ranges of possible *SELs* with centers in the unit square.

These two requirements lead to a scaling from the SNL radius, r_{SNL}, to a standard circle packing radius, r_{pack}:

$$r_{pack} = r/(2(1+r))$$

This yields the following process to determine the maximal (or upper bound on the) number of leaders (clusters) possible for a given radius, r:

1. Determine upper bound for number of leaders:

1.a Compute $r_{pack} = r/(2(1+r))$.

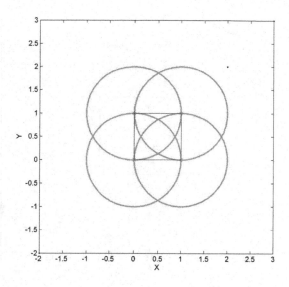

Figure 3.7: Maximal Packing of Leader *SEL*s in Unit Square.

1.b Find where r_{pack} falls in the Best Known Packing Results Table.

See Figure 3.8 for the transformation of the circle packing version of the 4 leaders.
Table 3.3 summarizes the results found for the set of radii considered previously:

Table 3.3. Upper Bound and SNL Average Cluster Size for Various Radii.

r_{SNL}	r_{pack}	Upper Bound	Average
1.000	0.2500	4	1.5
0.707	0.2071	6	2.5
0.500	0.1667	10	4.0
0.250	0.1000	25	12.0
0.100	0.0455	129	70.0
0.050	0.0238	1,849	250.0
0.010	0.0050	41,209	?4,500.0

Of course, it would also be interesting to find a leadership protocol that was equivalent to covering the unit square (see [117]) since this would require the minimum number
of leaders, but at the moment, this seems to be a complex computation; moreover, this
would greatly reduce cluster overlap.

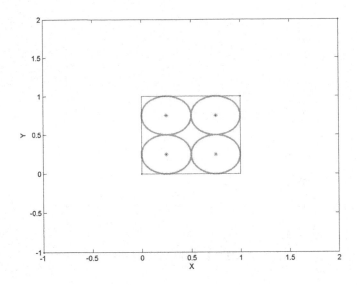

Figure 3.8: Maximal Packing of Leader *SEL*s in Unit Square.

3.3.5   Irregular Broadcast Region Shape

The results given previously assume a circular broadcast area, centered at the *SEL*. Ganesan has shown[43] that physical motes do not broadcast this way. Thus, we must examine how irregular broadcast shape influences the statistics determined above.

Using the data given by Ganesan et al. as the basis for a broadcast shape, the statistics for mean number of clusters was recomputed. Figure 3.9 shows the shape used as an approximation of the Berkeley mote's broadcast shape. A 271x336 array holds the characteristic function of the shape (i.e., 1 where the shape is, and 0 otherwise). These are scaled by 0.0194 in order to obtain a 5.2644x6.5270 unit rectangle so that the shape has area 4π (equivalent area to a circle with radius 2). Two *SEL*s are broadcast neighbors if the broadcast shape of each overlaps the location of the other. The orientations of these broadcast shapes are random across the *SEL*s.

Figure 3.10 shows the mean number of clusters for various numbers of motes randomly distributed in a 6x6 square. As can be seen, the average number of clusters approaches 8 as N grows larger.

3.4   Implementation

Next, we describe two complementary implementations of the SNL protocol: (1) on a set of Berkeley motes comprised of low-power 8-bit, 128Kb memory processors, communication devices and sensors, and (2) on a set of JStamps having 32-bit controllers, 2Mb of memory and native execution Java hardware.

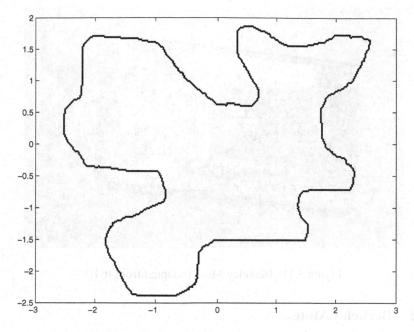

Figure 3.9: Approximation of Berkeley Mote Broadcast Shape (adapted from [58]).

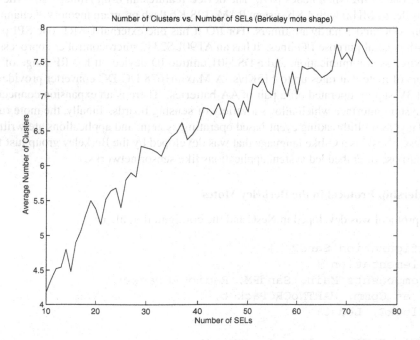

Figure 3.10: Average Number of Clusters vs. Number of *SEL*s in Network (adapted from [58]).

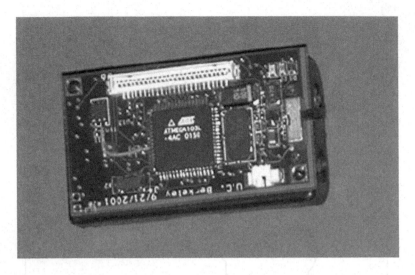

Figure 3.11: Berkeley Mote (adapted from [64]).

3.4.1   Berkeley Motes

We have developed one implementation in a set of four Berkeley motes. Figure 3.11 shows one of the Mica nodes [69]. The device features an 8-bit Atmega 103 Microcontroller (4 MHz) with 4 Kb system RAM, 128 Kb flash program memory, 8 channel, 10-bit ADC and 3 hardware timers. For I/O it has one external UART, one SPI port and 48 general purpose I/O lines. It has an AT90LS2343 microcontroller coprocessor for wireless communication, and a DS2401 unique ID device. It has RF range of up to tens of meters at rates up to 115Kb/s. A Maxim1678 DC-DC converter provides a solid 3V supply operated off a pair of AA batteries. There is an expansion connector I/O system interface which allows a variety of sensing boards. Finally, the mote runs the TinyOS multithreading event-based operating system, and applications are written in NesC; NesC is a C-like language that was developed by the Berkeley group just for the purpose of embedded system applications like sensor networks.

Leadership Protocol in the Berkeley Motes

The protocol was developed in NesC and the configuration file is:

```
configuration SandR {}
implementation {
   components Main, SandRM, RadioCRCPacket
      as Comm, UARTNoCRCPacket,
   ClockC, LedsC;

   Main.StdControl -> SandRM;

SandRM.UARTControl-> UARTNoCRCPacket;
```

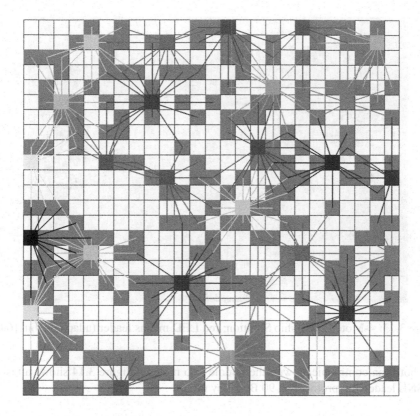

Figure 3.12: 250 Mote Leadership Solution from Mote Simulator (adapted from [64]).

```
SandRM.UARTSend-> UARTNoCRCPacket;
SandRM.UARTReceive-> UARTNoCRCPacket;

  SandRM.RadioControl -> Comm;
  SandRM.RadioSend -> Comm;
  SandRM.RadioReceive -> Comm;

  SandRM.Clock -> ClockC;
  SandRM.Leds -> LedsC;
}
```

The code was developed first in the Mote simulator, and Figure 3.12 shows a 250-node leadership solution. The gray squares have devices and the variable gray level squares are leaders. The edges show communication connectivity.

In the mote implementation, the leadership code takes 14.3Kb memory. A delay of 2 seconds is set for Phase I to allow neighbors lists to be built. Figure 3.13 shows four motes which have run the protocol; leader motes have the red LED illuminated. (The leader motes are the left and right motes which are not in each others broadcast range;

Figure 3.13: 4-Mote Leadership Solution; red LED means leader (adapted from [64]).

they both can communicate with the middle two motes.) Figure 3.14 shows a test of
the SNL leadership protocol on 90 Berkeley motes.

3.4.2   JStamp Processors

We have also implemented the S-Net algorithms in Systronix JStamps (see Figure 3.15).
There are many benefits to using Java as the programming language, and the JStamp or
JStik as the controller hardware. JStamp and JStik are physically small (JStamp is only
1x2 inches), yet contain a 32-bit controller, 2 Mbytes of memory, and the rich constructs
of Java. Software can be developed in Java on PCs and then easily loaded onto the
nodes. Another huge benefit of Java is the robust and proven security models designed
into the Java language and JXTA. Native execution Java hardware is physically small,
very power efficient, and computationally powerful. For example, the 1x2 inch JStamp
can run off a standard 9V transistor battery for up to 40 hours, and execute three million
Java byte codes per second. Systronix is currently the world leader in the commercial
development of such modules.

Of course, sensor networks do not always require wireless connectivity, and our
current JStamp testbed is set up as shown in Figure 3.16. Each JStamp in the testbed
has an RS232 connection to a PC, and the PCs are connected through Ethernet. (If we
use JStiks instead of JStamps, they have their own Ethernet ports and eliminate the need
for PCs. RF capability for JStamps/JStiks is also under development by Systronix.)

Independent processes are run on each PS which handle the communication between
JStamps; these processes connect to each other through sockets. The S-Net leadership
protocol and coordinate frame algorithm have been implemented in the JStamp testbed

Figure 3.14: Test of SNL Leadership Protocol with 90 Berkeley Motes.

Figure 3.15: Systronix JStamp Processor (adapted from [64]).

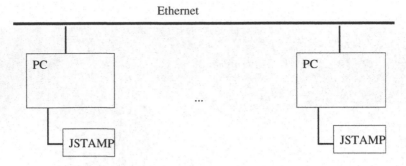

Figure 3.16: JStamp Testbed Layout (adapted from [64]).

with no problems encountered. There is an effect in setting timer values in the leadership protocol which is a critical issue in energy awareness in S-Nets.

3.5   Summary and Conclusions

These initial results of actual implementations of the S-Net algorithms are very encouraging. As pointed out by Chan and Perrig [17], the leadership protocol algorithm is the basis for many efficient wireless sensor network algorithms, including query processing [36], data aggregation [53, 165], routing [96, 155] and reliable broadcast [115, 154].

As far as comparing the two implementation testbeds, they have very complementary features. First, the Berkeley motes offer:

- small size

- low cost

- low power

- RF

- simulation environment

Mote cons include:

- small memory

- new programming language (NesC)

- differences between simulator and mote codes

- difficult to debug motes

The major issue in learning NesC is getting the communications aspects correct. In addition, there are some problems with shoehorning codes into the simulator (specified node connections may not occur in the simulator). In the actual motes, new batteries need to be used for benchmarking and testing to get consistent results. Moreover, the clock setting influences the correctness of the leadership protocol: set to 32 ticks/sec is really good; 64 ticks/sec results in failure about half the time, and 100 ticks/sec leads to high failure rates. In addition, delay timings are crucial for Phase I of the leadership protocol. Finally, simple acknowledgments in the frame algorithm led to more accurate results (angles between devices, etc.).

The JStamp testbed offers:

- Java programming

- off-stamp debugging

- small size

- low power

- large memory

- permits large memory sensors (e.g., CMUCam).

JStamp cons are:

- no RF

- no simulator for testbed

We have also explored parallel programming versions of SNL on multi-processor systems[55]. Simulations based on Unix processes, as well as MPI versions have been demonstrated and exploited.

To answer the question: "Does the SNL algorithm work as expected"? we have the following information. There is no mathematical proof at this time. However, it has been shown to work in the following cases:

- For all odd-sided (4-neighbor) regular grids from 3x3 up to 21x21.

- For n evenly spaced points on a circle (2 neighbors each) for n ranging from 4 to 200.

- For thousands of randomly generated graphs ranging in size from 10 to 100 nodes, and with average degree from 1 to 30.

Chapter 4

Coördinate Frames and Gradient Calculation

Computational Sensor Networks[1] depend on phenomenological models which describe spatio-temporal relations between physical quantities. This generally requires a common coördinate frame of reference. Almost all calculation depends on functions defined with respect to x, y, z, and t (e.g., the heat equation relates the partial derivative of temperature with respect to time to the second derivative of temperature with respect to space). Other quantities of interest, such as velocity, acceleration, momentum, etc., all depend on a frame of reference.

Generally, such a frame is assumed known or given; however, this is not usually the case for *SELs* in an *S-Net*. Such nodes are typically restricted in terms of hardware due to energy concerns and are used in places where GPS is not available (in buildings, cities, forests, etc.). In many cases it is sufficient to construct a local coördinate frame, that is, one given in terms of the *SELs* which define it; this is also called a relative frame. Of course, such a relative frame may include a large number of spatially distributed *SELs* so long as the necessary conditions hold (see below). It is also possible to anchor *SELs* to a global, or absolute, frame, like that provided by GPS, if necessary, and given enough global information.

This chapter provides the necessary tools to construct coördinate frames, and to relate them to mobile agents who desire to exploit them for localization and navigation.

4.1 Local and Global Coördinate Frames

To utilize the sensor population for localization and navigation purposes, local and/or global frames of reference need to be established. The assumption here is that the only information each sensor knows about the other sensors is their distance from it. Sensors can have uniquely identifiable tagged chirps, and can broadcast a signal that other sensors hear and relay back after some delay. Other physical phenomena may

---

[1]This chapter is a modified version of work done with Mohamed Dekhil, Scott Morris, Yu Chen and William B. Thompson [62], and Eddie Grant [63].

T.C. Henderson, *Computational Sensor Networks*, DOI: 10.1007/978-0-387-09643-8_4,
© Springer Science+Business Media, LLC 2009

be modeled and used to determine *SEL* location as well; e.g., the heat equation (see Chapters 8 and 9). Thus, a set of distances can be determined and from these an algorithm can compute positions with respect to a local sensor frame. The following is a formulation of this problem and its solution[2].

Problem: Given a set of n points, where $n \geq 3$, and the distances between each pair of points, $\{d_{ij}\}$, where $i = 1..n, j = 1..n$, and $i \neq j$, it is required to establish a planar frame of reference F where point i can be represented by the coördinates (x_i, y_i), where $i = 1..n$.

First we show how to incorporate unknown points into a known frame, and then show how to construct a local frame in which the frame origin is selected either arbitrarily or relative to a certain event or landmark.

4.1.1 Incorporating Points into a Coördinate Frame

In this formulation we show how knowledge of the coördinates of at least three points, p_1, p_2, and p_3, not laying on a straight line may be used to incorporate other points into a frame.

Assume that the coördinates of the first three points are: (x_1, y_1), (x_2, y_2), and (x_3, y_3), respectively, and the distances between these points and unknown point p_i are d_{i1}, d_{i2}, and d_{i3}, respectively, where $i = 4..n$ (see Figure 4.1). The distance can be expressed in terms of the unknown coördinates of point i, (x_i, y_i), as follows:

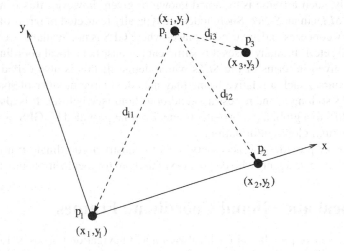

Figure 4.1: The Incorporation of at Point into a Known Frame (adapted from [62]).

---

[2]Note that if nodes have GPS or some other method to determine absolute location, then that obviates the determination of a relative frame.

$$(x_i - x_1)^2 + (y_i - y_1)^2 = d_{i1}^2 \qquad (4.1)$$
$$(x_i - x_2)^2 + (y_i - y_2)^2 = d_{i2}^2 \qquad (4.2)$$
$$(x_i - x_3)^2 + (y_i - y_3)^2 = d_{i3}^2 \qquad (4.3)$$

By subtracting Equation 4.2 from Equation 4.1 we get:

$$-2x_1x_i + x_1^2 + 2x_2x_i - x_2^2 - 2y_1y_i + y_1^2 + 2y_2y_i - y_2^2 = d_{i1}^2 - d_{i2}^2$$

and subtracting Equation 4.3 from Equation 4.1 yields:

$$-2x_1x_i + x_1^2 + 2x_3x_i - x_3^2 - 2y_1y_i + y_1^2 + 2y_3y_i - y_3^2 = d_{i1}^2 - d_{i3}^2$$

Simplifying the last two equations yields:

$$(x_2 - x_1)x_i + (y_2 - y_1)y_i = C_1 \qquad (4.4)$$

and

$$(x_3 - x_1)x_i + (y_3 - y_1)y_i = C_2 \qquad (4.5)$$

where

$$C_1 = \frac{1}{2}(d_{i1}^2 - d_{i2}^2 - x_1^2 + x_2^2 - y_1^2 + y_2^2)$$

and

$$C_2 = \frac{1}{2}(d_{i1}^2 - d_{i3}^2 - x_1^2 + x_3^2 - y_1^2 + y_3^2)$$

From Equation 4.4 we get:

$$x_i = \frac{C_1 - (y_2 - y_1)y_i}{(x_2 - x_1)} \qquad (4.6)$$

Substituting 4.6 into 4.5 we can calculate y_i as:

$$y_i = \frac{C_2 - \frac{(x_3 - x_1)}{(x_2 - x_1)}C_1}{(y_3 - y_1) - \frac{(x_3 - x_1)}{(x_2 - x_1)}(y_2 - y_1)}$$

Simplifying this equation yields:

$$y_i = \frac{(x_1 - x_3)C_1 + (x_2 - x_1)C_2}{y_1(x_3 - x_2) + y_2(x_1 - x_3) + y_3(x_2 - x_1)} \qquad (4.7)$$

Substituting for y_i in Equation 4.6 we get:

$$x_i = \frac{C_1 - (y_2 - y_1)\frac{(x_1 - x_3)C_1 + (x_2 - x_1)C_2}{y_1(x_3 - x_2) + y_2(x_1 - x_3) + y_3(x_2 - x_1)}}{(x_2 - x_1)}$$

Simplifying this last equation we get the solution for x_i as:

$$x_i = \frac{(y_3 - y_1)C_1 + (y_1 - y_2)C_2}{y_1(x_3 - x_2) + y_2(x_1 - x_3) + y_3(x_2 - x_1)} \qquad (4.8)$$

This formulation shows that the knowledge of the coördinates of 3 points not forming a straight line is sufficient to construct a frame and calculate the location of the other points with respect to that frame. The condition that the 3 points should not be on a straight line is necessary, otherwise, the denominator in Equations 4.7 and 4.8 will be zero.

4.1.2   Constructing a Local Frame

The main goal of the *S-Net* paradigm is to provide sensory information within a wide spatial area where clusters of small simple sensors are used to identify and locate certain events or actions in the environment. In many cases it is sufficient to use a local (or relative) frame to navigate and locate local events within the cluster range.

A local frame can be viewed as a special case of the global frame, where the position and orientation of the frame can be chosen arbitrarily. We now show how to construct a local frame for a set of n points, where $n \geq 3$.

1. Select any 3 points p_1, p_2, and p_3 not laying on a straight line; the set $\{p_1, p_2, p_3\}$ is called the *frame kernel*. Assume that the distances between these points are d_{12}, d_{13}, and d_{23}. The condition for non linearity is that the distance between any two points should be less than the sum of the distances between these two points and the third point:

$$d_{ij} < d_{ik} + d_{jk}$$

2. Set p_1 as the frame origin (i.e., $p_1 = (0,0)$).

3. Form the x-axis of the local frame as $\overline{p_1 p_2}$, where p_2 is constrained by a circle centered at p_1 with radius d_{12}. This means that $p_2 = (d_{12}, 0)$.

4. Calculate the location of the third point by selecting one of the two intersection points of two circles centered at p_1 and p_2 with radii d_{13} and d_{23}, respectively (see Figure 4.2). To calculate the location of p_3 we solve the following two equations:

$$(x_3 - x_1)^2 + (y_3 - y_1)^2 = d_{13}^2$$

$$(x_3 - x_2)^2 + (y_3 - y_2)^2 = d_{23}^2$$

Substituting the values of x_1, y_1, x_2, and y_2 from steps 2 and 3 above we get:

$$x_3^2 + y_3^2 = d_{13}^2$$

$$(x_3 - d_{12})^2 + y_3^2 = d_{23}^2$$

By solving these two equations for x_3 and y_3 we get:

$$x_3^2 = \frac{d_{12}^2 + d_{13}^2 - d_{23}^2}{2d_{12}} \tag{4.9}$$

$$y_3 = \pm \sqrt{d_{13}^2 - \frac{d_{12}^2 + d_{13}^2 - d_{23}^2}{2d_{12}}} \tag{4.10}$$

We then select one of the two locations for p_3. Here we select the positive value of y_3.

5. Use the results to incorporate points into a known frame to get the (x, y) locations of the remaining points with respect to this local frame.

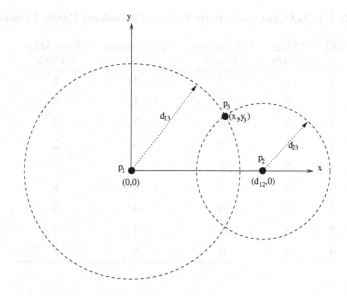

Figure 4.2: Construction of a Local Frame (adapted from [62]).

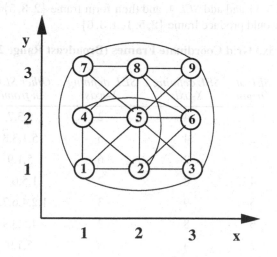

Figure 4.3: 3x3 Grid.

We have developed a Matlab code which calculates a coördinate frame for each *SEL* whenever that is possible in terms of its neighbors. For the 3x3 grid shown in Figure 4.3, the local frames are determined as shown in Table 4-1.

As can be seen, no *SEL* succeeds in determining the coördinates of more than four points. This is due to the fact that the distance from three established *SEL*s must be known to a new point and usually only a couple of distances are known to points already in the frame. This fails for more points because the broadcast range is 1.5 units. If this is raised to 2 units, we get the result shown in Table 4-2. (Note that is is

Table 4-1. 3x3 Grid Coördinate Frames (Broadcast Range 2 Units).

SEL	SEL at Origin	SEL defining X-axis	SEL defining Y-axis	Other SELs in frame
1	1	2	4	5
2	2	1	5	4
3	3	6	2	5
4	4	1	5	2
5	5	6	2	3
6	6	5	9	8
7	7	8	4	5
8	8	5	7	4
9	9	6	8	5

possible to produce multiple frames at a *SEL* and then merge them; e.g., *SEL* 2 could produce frame $\{2, 5, 1\}$ and add *SEL* 4, and then form frame $\{2, 3, 5\}$ and add *SEL* 6. Combining these would produce frame $\{2, 5, 1, 4, 3, 6\}$.

Table 4-2. 3x3 Grid Coördinate Frames (Broadcast Range 2 Units).

SEL	SEL at Origin	SEL defining X-axis	SEL defining Y-axis	Other SELs in frame
1	1	4	2	5,3,7
2	2	4	6	5,1,3,8
3	3	2	6	5,1,9
4	4	8	2	1,5,6,7
5	5	9	3	1,2,4,6,7,8
6	6	8	2	4,5,3,9
7	7	4	8	5,1,9
8	8	4	6	2,5,7,9
9	9	8	6	5,3,7

4.1.3   Moving between Local Frames

Moving from one sensor cluster to another requires a transformation between the two local frames. To determine the transformation functions, there must be at least two points common to both frames. Using a homogeneous transformation, the relationship between the (x, y) locations of a point p_i with respect to both frames can be written as (see Figure 4.4):

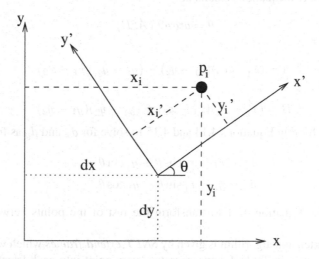

Figure 4.4: Transformation between Frames (adapted from [62]).

$$\begin{bmatrix} x_i \\ y_i \\ 1 \end{bmatrix} = \begin{bmatrix} \cos\theta & -\sin\theta & d_x \\ \sin\theta & \cos\theta & d_y \\ 0 & 0 & 1 \end{bmatrix} \begin{bmatrix} x_i' \\ y_i' \\ 1 \end{bmatrix} \qquad (4.11)$$

Thus, we have two equations in three unknowns; d_x, d_y, and θ.

$$x_i = x_i' \cos\theta - y_i' \sin\theta + d_x \qquad (4.12)$$

$$y_i = x_i' \sin\theta + y_i' \cos\theta + d_y \qquad (4.13)$$

Therefore, we need at least two common points, p_1 and p_2, to solve Equations 4.12 and 4.13.

Assume that we have two common points p_1 and p_2 with coördinates (x_1, y_1) and (x_2, y_2) with respect to the first frame, and (x_1', y_1') and (x_2', y_2') with respect to the second frame. Substituting these points in Equations 4.12 and 4.13 yields:

$$x_1 = x_1' \cos\theta - y_1' \sin\theta + d_x \qquad (4.14)$$
$$y_1 = x_1' \sin\theta + y_1' \cos\theta + d_y \qquad (4.15)$$
$$x_2 = x_2' \cos\theta - y_2' \sin\theta + d_x \qquad (4.16)$$
$$y_2 = x_2' \sin\theta + y_2' \cos\theta + d_y \qquad (4.17)$$

By subtracting Equation 4.16 from Equation 4.14 and Equation 4.17 from Equation 4.15 we get:

$$(x_1 - x_2) = (x_1' - x_2')\cos\theta - (y_1' - y_2')\sin\theta \qquad (4.18)$$
$$(y_1 - y_2) = (x_1' - x_2')\sin\theta + (y_1' - y_2')\cos\theta \qquad (4.19)$$

Solving the last two equations results in:

$$\theta = atan2\,(A, B) \qquad (4.20)$$

where

$$A = (x_1' - x_2')(y_1 - y_2) - (y_1' - y_2')(x_1 - x_2)$$

and

$$B = (x_1' - x_2')(x_1 - x_2) + (y_1' - y_2')(y_1 - y_2)$$

Substitute for θ in Equations 4.14 and 4.15 to solve for d_x and d_y as follows:

$$d_x = x_1 - x_1'\cos\theta + y_1'\sin\theta \qquad (4.21)$$
$$d_y = y_1 - x_1'\sin\theta - y_1'\cos\theta \qquad (4.22)$$

Finally, we use Equation 4.11 to transform the rest of the points between the two frames.

A frame extension algorithm is given by *SNET_extend_frames* which when applied to the frames given in Table 4-1 incorporates every point into each frame. Note that when mapping points from one frame to another, it is necessary to make sure that the frames have the same sense (i.e., both right-handed or left-handed). They must have three common points for this check.

We have developed algorithms to compute a coördinate frame for a cluster of Berkeley motes. Figure 4.5 shows a 250-mote simulation result with the local frames shown, and Figure 4.6 shows the local frame neighborhoods. We have run the code on the 4 Berkeley motes and produced correct frames for them as well; the coordinate frame executable takes 21 Kb. The leadership and coordinate frame executable takes 133.4 Kb memory.

Exploiting sensor networks involves understanding algorithmic and engineering issues of real-world devices, and making both raw and processed data readily accessible to humans. We have implemented these algorithms on two complementary domains: Berkeley motes and JStamp embedded processors [64].

4.2   Gradient Calculation

Given a coördinate frame and a set of samples from a real-valued function at various points in the frame, the partial derivatives of the function in X and Y are generally of interest, as is the vector formed from these derivatives, the gradient. Much is known analytically about derivatives, and the approximation of derivatives from sampled data, but this is usually performed on a regular grid with equi-spaced points. With Computational Sensor Networks, a major issue is that *SELs* are not generally deployed in a

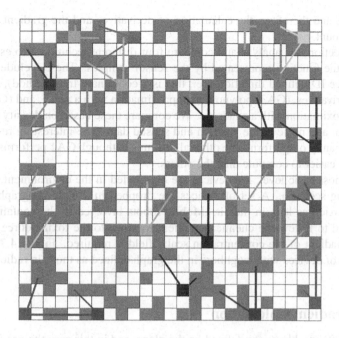

Figure 4.5: 250-Mote Coordinate Frames Calculation; mote simulation (adapted from [64]).

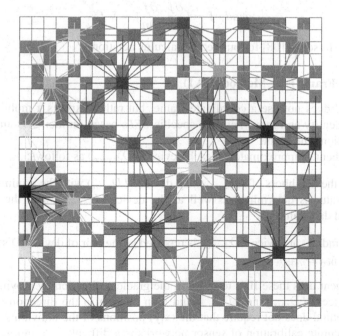

Figure 4.6: A Selection of S-Element Devices and their Local Neighbors (adapted from [64]).

regular grid, and therefore, the calculation of derivatives and the gradient must take this into account.

In this section, we analyze and compare four different techniques to estimate the gradient of the function represented by the sensor samples. These include: (GA1) a simple device ID defined direction (i.e., this is a coördinate-free method), (GA2) directional derivatives, (GA3) polynomial approximation with a plane, and (GA4) polynomial approximation with a quadratic. We compare these based on density of devices per unit area, and noise in the position and sensed data. The interesting result is that GA3 significantly outperforms the other algorithms, although GA1 performs very well and is much easier to compute than the others.

At the most basic level, the devices are distributed in the environment. Consider the following scenario. A set of devices are dropped in a wide geographic area to monitor a toxic gas leak in the air. Mobile robots involved in the containment and cleanup need to follow the chemical gradient to move to the toxin source locations. Thus, the gradient of the concentration scalar field is required. Figure 4.7 shows an example set of devices with neighbors in the graph defined as those in radio broadcast range.

4.2.1   Gradient Calculation

Generally, CSN problems are defined on the plane, and in this case the *gradient* of f at (x, y) is the vector in $\Re^2$ given by:

$$grad \ \ f = (\frac{\partial f}{\partial x}, \frac{\partial f}{\partial y}) = \nabla f$$

(We follow Marsden [108] for our vector calculus notation.)

Coördinate Frame Free Method (GA1)

This method does not use a coördinate frame, but rather expects a mobile agent to be able to identify and move between specific *SEL*s based on their IDs and wireless communication with them.

The gradient is approximated at each device (ID_{device}) as follows:

1. From the neighbors of ID_{device}, the device (ID_{min}) with the minimum sensed data value is determined as well as the device (ID_{max}) with the maximum sensed data value.

2. The gradient at device ID_{device} is reported as the pair of device ID's: (ID_{min},ID_{max}).

A mobile agent uses these ID's to move in the gradient direction by moving between the two devices, ID_{min} and ID_{max}, and then moving in the direction of ID_{max}. Note the method does not require that the (x,y) positions of the individual devices be known. Accurate calibration of sensor networks is a difficult task and getting good position data can be very difficult or very expensive [161], thus making this approach useful when no coördinate frame is available. This method is very inexpensive and robust and thus, very attractive if its performance compares well to the other more rigorous approaches.

Directional Derivatives (GA2)

This method requires knowledge of the positions of the devices. The *directional derivative* of f at $\overline{x}$ in direction $\overline{v}$ is given by:

$$d.d. = \frac{d}{dt} f(\overline{x} + t\overline{v}) \mid_{t=0}$$

if it exists. From this, we have that the directional derivative is also defined by:

$$d.d. = \lim_{h \to 0} \frac{f(\overline{x} + h\overline{v}) - f(\overline{x})}{h}$$

We approximate this by:

$$d.d.a. = \frac{f(\overline{x} + h\overline{v}) - f(\overline{x})}{h} \tag{4.23}$$

Assuming all directional derivatives exist, it is the case that:

$$\overline{D}f(\overline{x})\overline{v} = grad\ f(\overline{x}) \cdot \overline{v} = \nabla f(\overline{x}) \cdot \overline{v}$$

so that:

$$d.d. = [\frac{\partial f}{\partial x}(\overline{x})]v_x + [\frac{\partial f}{\partial y}(\overline{x})]v_y \tag{4.24}$$

where $\overline{v} = (v_x, v_y)$. Combining these, we approximate the gradient at each device, ID_{device}, located at $\overline{e_0}$, as follows:

1. Choose two of ID_{device}'s neighbors, ID_1 and ID_2, located at $\overline{e_1}$ and $\overline{e_2}$, respectively, such that $\angle \overline{e_1 e_0 e_2}$ is as close to a right angle as possible.

2. For the two points, $\overline{e_1}$ and $\overline{e_2}$, solve (4.23) to get the following pair of equations:

$$d.d.a._1 = f_x e_{1x} + f_y e_{1y} \tag{4.25}$$

$$d.d.a._2 = f_x e_{2x} + f_y e_{2y} \tag{4.26}$$

3. Solve (4.25) and (4.26) for the two unknowns: f_x and f_y and form the gradient as (f_x, f_y).

Polynomial Approximation: Plane (GA3)

For each device, the position must be known. To approximate the gradient:

1. From the positions and sensed data values of the n points within broadcast range of the current device (i.e., itself and its neighbors), form the linear system:

$$\begin{pmatrix} f(\overline{e_1}) \\ f(\overline{e_2}) \\ \vdots \\ f(\overline{e_n}) \end{pmatrix} = \begin{bmatrix} 1 & e_{1x} & e_{1y} \\ 1 & e_{2x} & e_{2y} \\ & \vdots & \\ 1 & e_{nx} & e_{ny} \end{bmatrix} \begin{pmatrix} a_0 \\ a_1 \\ a_2 \end{pmatrix} \tag{4.27}$$

2. Solve (4.27) for a_0, a_1, and a_2.

3. The gradient is then (a_1, a_2).

Polynomial Approximation: Quadratic (GA4)

Here we make the assumption that the functional form of the sensed data is:

$$f(x,y) = \frac{D_{max}}{\sqrt{1 + (S_x - x)^2 + (S_y - y)^2}} \tag{4.28}$$

where D_{max} is the maximum value of the function at the source location (S_x, S_y). In order to set up to solve for the gradient, rewrite (4.28) as follows:

$$\frac{1}{f(x,y)} = \frac{\sqrt{1 + (S_x - x)^2 + (S_y - y)^2}}{D_{max}}$$

$$\frac{1}{f^2(x,y)} = \frac{1 + (S_x - x)^2 + (S_y - y)^2}{D_{max}^2}$$

$$u(x,y) = a_0 + a_1 x + a_2 y + a_3 x^2 + a_4 y^2$$

where $u(x,y) = \frac{1}{f^2(x,y)}$, $a_0 = \frac{1 + S_x^2 + S_y^2}{D_{max}^2}$, $a_1 = \frac{-2S_x}{D_{max}^2}$, $a_2 = \frac{-2S_y}{D_{max}^2}$, $a_3 = \frac{1}{D_{max}^2}$, and $a_4 = \frac{1}{D_{max}^2}$.

Then the gradient is found as:

1. From the positions and sensed data of the device and its neighbors, form the linear system:

$$\begin{pmatrix} u(\overline{e_1}) \\ u(\overline{e_2}) \\ \vdots \\ u(\overline{e_n}) \end{pmatrix} = \begin{bmatrix} 1 & e_{1x} & e_{1y} & e_{1x}^2 & e_{1y}^2 \\ 1 & e_{2x} & e_{2y} & e_{2x}^2 & e_{2y}^2 \\ & & \vdots & & \\ 1 & e_{nx} & e_{ny} & e_{nx}^2 & e_{ny}^2 \end{bmatrix} \begin{pmatrix} a_0 \\ a_1 \\ a_2 \\ a_3 \\ a_4 \end{pmatrix} \tag{4.29}$$

2. Solve (4.29) for a_0, a_1, a_2, a_3, and a_4.

3. Then the gradient is given by $(2a_3 x + a_1, 2a_4 y + a_2)$.

Note that $S_x = \frac{-a_1}{2a_4}$, $S_y = \frac{-a_2}{2a_4}$, and $D_{max} = \frac{1}{\sqrt{a_4}}$. Note that the recovery of these parameters is difficult as the a_i's are very sensitive to the data.

4.2.2 Simulation Experiments

Our simulation works as follows:

for number of devices = dev_{min} to dev_{max}

    for noise = 0 to $noise_{max}$

        for number of trials = 1 to $trials_{max}$

            Distribute devices uniformly over area

            Select source location for scalar field

            Set values of sensor devices

            for each gradient method

                Calculate the gradient

                Calculate the error

            end

        end

    end

end

This has been implemented in Matlab, and Table 4-3. gives the results of the simulations.

The density of the devices was allowed to vary from 1 per unit area to 10 per unit area. The table shows that increasing the number of devices generally improves the quality of the approximation. Two noise levels were evaluated: (1) no noise, and (2) noise with standard deviation 1. This noise is applied to both the position of the devices, as well as to the sensed data values. That is, the device position is normally distributed about the actual position with 0 mean and either standard deviation of 0 or 1. Sensed data is handled in a similar manner.

4.2.3 Conclusion

From Table 4-3 it can be seen that GA3 performs significantly better than the other algorithms, even under noisy conditions. Figure 4.7 shows a sample sensor network with neighbors graph, and Figure 4.8 shows the gradient approximation by GA3 with an average of 2 devices per unit area and no noise. Moreover, GA1 - which does not use device position information - performs comparable to the other algorithms, and in absolute terms is not so bad (about 16 degrees error under noisy conditions with a couple of devices per unit area). Figure 4.9 shows the results under the same conditions as GA3 above.

Figures 4.10 and 4.11 show the results of GA2 and GA4 on the same data. As can be seen, GA4 does a very poor job of approximating the gradient; any time a specific functional form is chosen, it will do poorly if it does not match the actual environment.

Table 4-3. Simulation Results.

Algorithm	Angle Error (degrees)	Angle Error Std	Mag Error (pixels)	Mag Error Std
100 devices / 0 data error				
GA1	5.89	0.02	5.85	1.49
GA2	5.39	0.02	10.71	8.40
GA3	1.20	0.01	2.90	0.62
GA4	16.63	0.10	10.70	19.12
200 devices / 0 data error				
GA1	2.92	0.01	5.34	1.34
GA2	6.18	0.01	14.52	21.42
GA3	0.67	0.01	2.61	0.51
GA4	14.83	0.10	6.90	3.54
100 devices / 1 std data error				
GA1	25.79	0.02	6.01	1.43
GA2	20.90	0.06	7.72	2.74
GA3	10.24	0.04	6.49	2.70
GA4	18.59	0.06	56.4	326.94
200 devices / 1 data error				
GA1	15.69	0.03	5.72	1.37
GA2	10.61	0.04	8.67	3.44
GA3	6.58	0.04	5.81	0.97
GA4	17.87	0.08	52.80	150.21

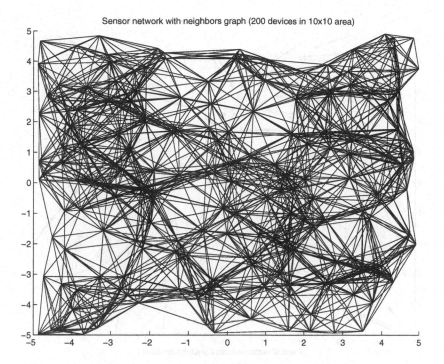

Figure 4.7: Sample Sensor Network with Neighbors Graph (adapted from [63]).

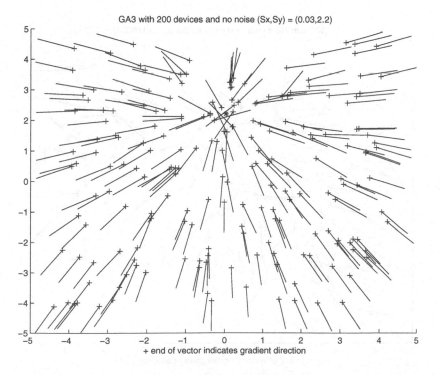

Figure 4.8: GA3 Gradient Approximation with No Noise (adapted from [63]).

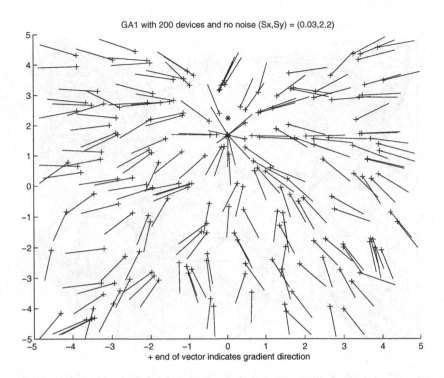

Figure 4.9: GA1 Gradient Approximation with No Noise (adapted from [63]).

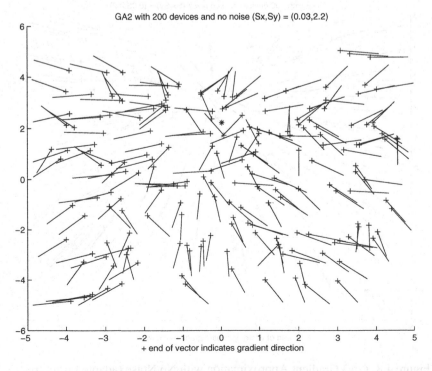

Figure 4.10: GA2 Gradient Approximation with No Noise (adapted from [63]).

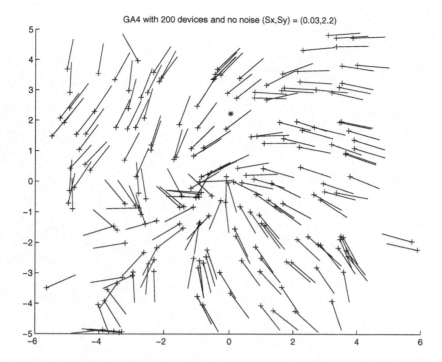

Figure 4.11: GA4 Gradient Approximation with No Noise (adapted from [63]).

Methods have been given for (1) the determination of local coördinate frames, given the pairwise distances between nodes, (2) the transformation between local frames which share at least two points, (3) several methods for the calculation of the gradient in a sensor network, including a coördinate free technique.

Figure 4.11: GM4 Gradient Approximation with No Noise (adapted from [31]).

Methods have been given for: (1) the determination of local coordinate frames, given the pairwise distances between nodes; (2) the transformation between local frames which share at least two points. These are the basis for the calculation of the gradient in a sensor network, including a coordinate-free technique.

Chapter 5

Pattern Formation in S-Nets

Biological systems exhibit an amazing array of distributed sensor/actuator systems, and the exploitation of principles and practices found in nature will lead to more effective artificial systems. The retina is an example of a highly tuned sensing organ, and the human skin is comprised of a set of heterogeneous sensor and actuator elements. Moreover, the specific organization and architecture of these systems depends on contextual influences during the developmental stages of the organism. Comparable theoretical and technological methodologies need to be found for wireless sensor networks. We propose the study of reaction-diffusion systems from mathematical biology as a starting point for this endeavor. Algorithms and experiments are described here for a useful set of pattern formation methods in wireless sensor networks.

Alan Turing introduced a revolutionary reaction-diffusion model as the chemical basis of morphogenesis [156], and this method lends itself particularly well to pattern synthesis in distributed systems. For more detailed explanations, see his original paper (which provides an exemplar of the scientific paper – theory, analysis and numerical solution on the Manchester machine which Turing helped design and build!), as well as the works of Murray [111], Meinhardt [109], and more recently, Maini and Othmer [106]. Turing's key insight was that diffusion of an inhibitory morphogen could lead to the formation of stable and variegated patterns. This is related to nonlinear far from equilibrium thermodynamics, and dissipative structures (e.g., see Prigogine [116, 123, 124] who received the Nobel prize in chemistry for work in this area). One goal of our work is to understand how these principles are at work in biological sensor systems and how they may be exploited in wireless sensor networks.

We have previously proposed to use Turing's reaction-diffusion mechanism to generate patterns in wireless sensor networks. [67]. The basis of this mechanism is a set of equations that capture the reaction and diffusion aspects of certain chemical kinetics:

$$\frac{\partial \mathbf{c}}{\partial t} = f(\mathbf{c}) + D\nabla^2 \mathbf{c} \tag{5.1}$$

where $f(\mathbf{c})$ describes the reaction and $D\nabla^2 c$ expresses the diffusion component. The simplest such systems have two *morphogens* or variables; one of these acts as the

T.C. Henderson, *Computational Sensor Networks*, DOI: 10.1007/978-0-387-09643-8_5,
© Springer Science+Business Media, LLC 2009

activator and the other acts as the inhibitor. The two variable system can be modeled
by:

$$\frac{\partial u}{\partial t} = \gamma f(u,v) + \nabla^2 u, \frac{\partial v}{\partial t} = \gamma g(u,v) + d\nabla^2 v \tag{5.2}$$

where u and v are the concentrations of the morphogens, d is the diffusion coefficient
and γ is a constant measure of scale. The functions $f(u,v)$ and $g(u,v)$ represent the
reaction kinetics. As an example, we have explored the generation of spatial patterns
using the Turing system of equations:

$$f(u,v) = \beta - uv, g(u,v) = uv - v - \alpha$$

where u and v are the morphogen concentrations, α and β are the decay and growth
rates, respectively, and γ sets the speed of the reaction. They define a domain in which
Equation (2) becomes linearly unstable to certain spatial disturbances. This domain
is referred to as *Turing space* where the concentrations of the two morphogens will
become unstable and result in the patterns shown in Figure 5.1.

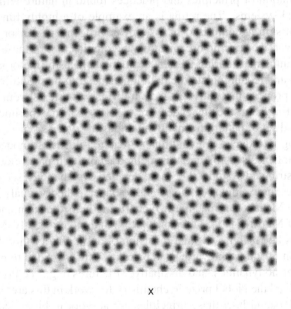

Figure 5.1: Turing Spot Pattern.

The pattern is the result of each cell running the equations locally while *diffusing*
to its neighbors; a stable solution may be thresholded to produce a binary value at each
sensor, and the total of these gives the pattern. Note that the distribution of these spots
is close to hexagonal.

We introduced the use of Turing's reaction-diffusion pattern formation to support
high-level tasks in sensor networks (*S-Nets*). This has led us to explore various biologically motivated mechanisms. We address below some issues that arise in trying to get
reliable, efficient patterns in irregular grids.

Much remains to be done at the higher level of information extraction, interpretation and exploitation of networked sensor systems. Our central thesis is that bio-based engineering will lead to strong solutions in this domain; that is, we propose to identify and ultimately incorporate effective computational strategies used by biological systems. The challenge is to identify mechanisms that lead to algorithms or paradigms that are reliable, inexpensive and ubiquitous in many applications.

Others have explored the use of both reaction-diffusion and more general diffusion methods in computer vision and robotics. For example, Fukuda et al. describe the use of reaction-diffusion techniques in robot motion[?]. Moreover, as described by Peronna et al.[?], multi-scale descriptions of images (i.e., scale-space) can be produced by embedding the original image in a family of images obtained by convolving the original image with a filter; Koenderink[?] showed that this is equivalent to finding the solution of the diffusion equation:

$$I_t = \nabla^2 I = I_{xx} + I_{yy}$$

We believe that it will be quite useful for *S-Nets* to use similar methods to analyze sensed data of various sorts. Other proposed diffusion models include, for example, [?] who proposes directed diffusion - a datacentric communication coördination technique that "enables energy savings by selecting empirically good paths and by caching and processing data in- network." The focus of such work is more on the networking and operating systems aspects of the sensor network, whereas our work is more concerned with the sensor network as a computation engine itself. More closely related to our work is that of Justh and Krishnaprasad [?] who propose the active coordination of a large array of microactuators by means of diffusive coupling implemented as interconnection templates, and Nagpal [?] who describes methods to create patterns of diverse geometry. We believe that this style of research will reap great benefits in three aspects: (1) network *morphogenesis*, (2) sensed data analysis, and (3) display pattern synthesis.

As Meinhardt points out [?], "the control of development in a higher organism is one of the major unresolved problems in biology ... in a developmental system a signaling and signal-receiving mechanism must exist which enables the cell to communicate in a manner appropriate to its position ... [the] goal is to show which interactions of substances can lead to such signaling systems and how the cells then can respond to these signals in order that stable states of determination are attained." This matches our view of the core issues, and we see that their solution can heavily impact sensor network algorithms as well.

For example, consider a forest fire scenario: sensor devices are dropped into a wide geographic area, establish a network, compute coördinate frames, calculate gradients, and produce a stripe pattern of off-on signals that can be used by fire fighting agents to go to a fire control point by following *on* devices (pattern == 1) and return by following *off* devices (pattern == 0) (see Figure 1.2). Such patterns can be computed by very robust reaction-diffusion systems derived from models of biological pattern formation.

Our general research program is to explore a small set of biological sensing and signaling mechanisms, and we hope to make significant contributions by providing

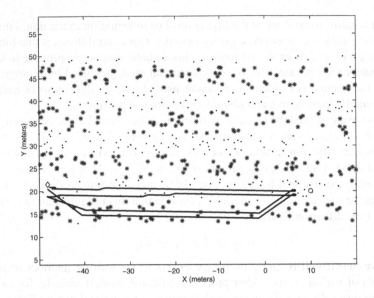

Figure 5.2: Robot Path in Reaction Diffusion Pattern (○ is the fire control point; ◇ is the robot load point).

(1) biologically realistic models and efficient computational counterparts, (2) fault tolerant frameworks in which to run them, and (3) demonstrations of their application in human interface and large-scale sensor networks. In addition, we are building *S-Net* simulation, emulation, and experimentation testbeds [64]. Here we describe some initial results in the first of these areas.

Patterns in the *S-Net* can be used to support many high-level algorithms or activities:

- stripe, spot or ring patterns can be used as encoders for physical or logical purposes; for example, a robot can keep track of how far it has traveled (physical), or communication packets can travel along certain stripes to minimize power cost or to avoid congestion (logical).

- certain sets of patterns form a basis set for 2D images (e.g., Haar or Hadamard basis sets); any map (topo, etc.) or image can then be encoded in terms of the coefficients associated with the respective basis images.

- the patterns can be used as a reference wave so that sensed data (or features derived from it) can be encoded as an interference pattern (i.e., a hologram)

- moving waves can also be computed, and thus the *S-Net* can serve as a signal carrier or modulator.

Understanding the precision and reliability of pattern formation is then of high importance.

The most common application of *S-Nets* is to serve as a sensory organ; e.g., to capture images, sounds, chemical concentrations, temperature, etc., over the region of interest. However, *S-Nets* may also be used as a display, either directly through LED's or by making values avaliable upon query. A biological analogy is the skin; a zebra's stripes provide coloration as well as a myriad of sensors embedded in the skin, and these serve to provide some ecological advantage. A combination of these is especially interesting where the display is influenced by the sensing; e.g., for camouflage effects like in the chameleon.

Given a set of *SELs* in the plane, it may be useful to store and exploit a pattern in the *S-Net*. For example, stripes may be useful for several purposes: (1) as pathways for physical or logical tasks, e.g., mobile agents trajectories or packet transmission, (2) as distance encoders as mobile agents cross them, or (3) as boundary markers. The ability to store arbitrary patterns (e.g., maps) has been shown useful [62, 19] for calculating shortest paths for robots to follow through terrain with varying traversability properties.

If the *SELs* are situated in a coördinate frame, i.e., each *SEL* knows its (x, y) location, then given a specification of the pattern as a function $f(x, y)$, each *SEL* can determine its own value. Useful binary patterns may also be represented as a bit stored at each *SEL*, which requires a thresholding function $t(x, y)$, as well; this works well for paths or checkerboards. Moreover, combinations of stripe and checkerboard patterns may serve as a set of 2D basis functions to allow the representation of any arbitrary pattern as a linear combination of them.

5.1 Regular Geometric Figures

Equi-spaced straight line segments, stripes, with orientation θ, may be easily computed by means of the following function ([111]):

$$f(x, y) = cos(x cos(\theta) + y sin(\theta))$$

Figure 5.3 shows the result with $\theta = 0$ and $\theta = \pi/4$.

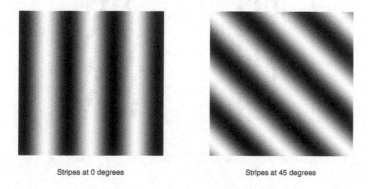

Stripes at 0 degrees   Stripes at 45 degrees

Figure 5.3: Vertical and 45° Stripes.

If the range of $f(x, y)$ is $[-1, 1]$, then a mobile agent can determine how far it is from the center of the nearest stripe. Consider now the case of a set of *SELs* randomly distributed in a square area (each x and y coördinate is sampled from a uniform distribution). Figure 5.4 shows an instance of 1,000 *SEL*s. Each *SEL* can determine its own value using the formula given above, and mobile agents can request this information in order to stay near the stripe of interest; in this way, the mobile agent does not need to be incorporated into the *S-Net* coördinate frame. Figure 5.5 shows the set of *SEL*s with values near 1 shown as a circle ('o') and values near -1 shown as a dot ('.') for both stripe orientation 0 and $\pi/4$.

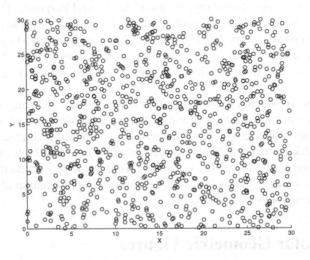

Figure 5.4: Sample of 1,000 *SEL*s in a Square Area.

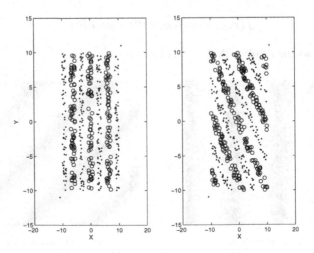

Figure 5.5: Vertical and 45° Stripes in *S-Net*.

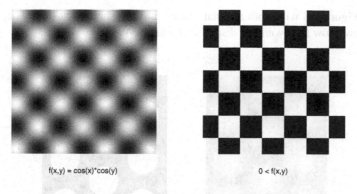

f(x,y) = cos(x)*cos(y)   0 < f(x,y)

Figure 5.6: Checkerboard in the Plane.

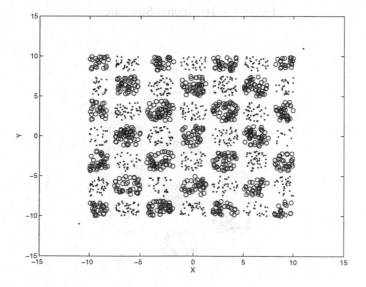

Figure 5.7: Checkerboard in *S-Net*.

Another useful pattern is the checkerboard. The function:

$$f(x, y) = cos(x)cos(y)$$

thresholded by $0 < f(x, y)$ yields a checkerboard oriented along the x and y axes (see Figure 5.6 for both the original and thresholded versions). Using 2,000 *SEL*s, it is possible to approximate the squares in an *S-Net* (thresholded at $0.3 < f(x, y)$ and $f(x, y) < -0.3$ as shown in Figure 5.7).

As a final example, consider the hexagonal structure defined by:

$$f(x, y) = cos(\frac{y\sqrt{3} + x}{2}) + cos(\frac{y\sqrt{3} - x}{2}) + cos(x)$$

shown in Figure 5.8 with both original values and thresholded at $0 < f(x, y)$. Figure 5.9 shows how this would be displayed by an *S-Net*.

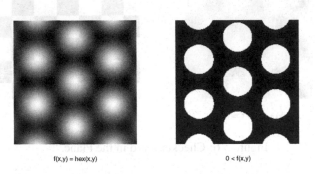

Figure 5.8: Hexagonal Structure.

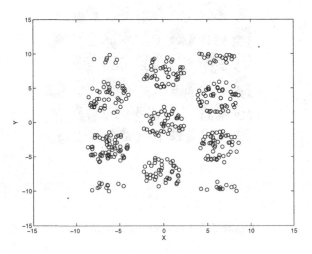

Figure 5.9: Hexagonal Structure in *S-Net*.

Of course, if coördinates are available, any pattern which can be expressed as a 2D function can be approximated by an *S-Net*. The quality of the approximation must be evaluated in probabilistic terms when the distribution of *SEL*s is random.

These regular patterns have been implemented on a set of Moteiv Tmote Sky motes, and the results are shown in Figures 5.10 through 5.14. Figure 5.10 shows the set of the 42 motes as laid out for these experiments. Figures 5.11 and 5.12 show the vertical and $45°$ lines, while Figures 5.13 and 5.14 show the checkerboard and hexagonal layouts, respectively.

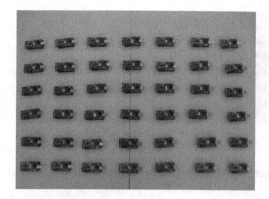

Figure 5.10: Layout of 42 X-bow T-Sky Motes.

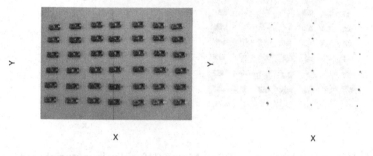

X   X

Figure 5.11: Vertical Lines in T-Sky Motes (left side: layout; right side: LED's turned on).

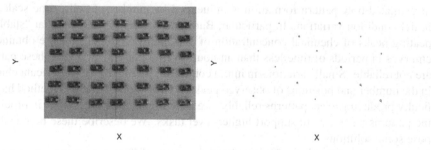

X   X

Figure 5.12: 45° Lines in T-Sky Motes (left side: layout; right side: LED's turned on).

5.2 Reaction-Diffusion Patterns

Some work has already been done to determine the range and type of patterns possible with the Turing pattern formation approach. Theoretical aspects have been studied and regions of the parameter space characterized as they relate to pattern formation (i.e., the parameters are the coefficients in the PDEs) [4, 46, 97]. Others have

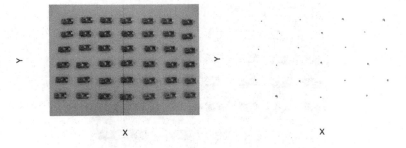

Figure 5.13: Checkerboard Layout in T-Sky Motes (left side: layout; right side: LED's turned on).

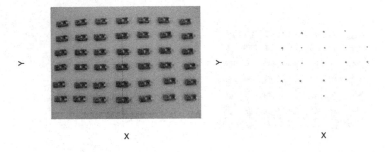

Figure 5.14: Hexagonal Layouts in T-Sky Motes (left side: layout; right side: LED's turned on).

investigated how pattern formation is influenced by number of cells, time scale, and initial condition variation. In particular, Bard and Lauder [7] showed that "stable repeating peaks of chemical concentration of periodicity 2-20 cells can be obtained in embryos in periods of time less than an hour. We do find however that these patterns are not reliable. Small variations in initial conditions give small but significant changes in the number and positions of observed peaks." They showed that this method has difficulty producing exact patterns reliably. We have found other difficulties in producing the patterns necessary to support higher-level tasks. We describe these here and propose some solutions.

A more significant issue for us is that the reaction-diffusion pattern formation equations assume that the inter-cell distance is uniform (and usually equal to 1). Our *S-Nets*, however, do not form a uniformly spaced grid in 1D or 2D; in fact, we generally assume that the sensor devices are randomly dropped in the environment. In addition, the diffusion part of the equations uses the inter-node distances in the computation of the second derivative. Two concerns are:

- these distances are not uniform, and
- in an actual implementation, there will be some amount of error in the inter-node distance determination.

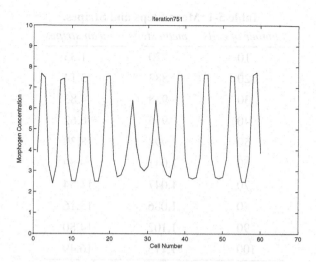

Figure 5.15: Typical 1-D Turing Pattern.

This has led us to investigate the impact of non-uniform spacing on the pattern computation.

The basic 1D Turing reaction-diffusion mechanism produces a pattern as shown in Figure 5.15, and takes about 1,040 iterations to converge for a set of 60 cells. A set of simulation experiments were run with from 10 to 100 cells in steps of 10. Table 5-1 gives the results for the mean number of steps to converge (total change in morphogen a is less than 0.0001), and the mean number of stripes formed.

Next consider what happens when non-unit distances are introduced as the inter-device distances. The point locations are determined as follows:

- start with n equi-spaced points, 1 unit length apart,

- add uniform noise to the location with max x and y distances from the grid positions varying from 0 to 0.65.

Table 5-2 gives the average number of failures to converge for the Turing pattern. This result, as well as similar results with 2D patterns, argues against using actual inter-SEL distances for the reaction-diffusion computation. However, this turns out to be an advantage since the inter-SEL distances may be difficult to determine. Moreover, the patterns will form based purely on the topological nature of the SEL interconnectivity graph. This can be very advantageous in 2D pattern formation in the *S-Net*.

Figure 5.1 shows the Turing pattern, but it is formed in an $n \times n$ array, where every array element has two vertical neighbors and two horizontal neighbors (boundary elements wrap around to the opposite boundary elements), and all neighbors are at unit distance. This makes the Laplacian calculation reasonably accurate.

As a first approximation to an arbitrary *S-Net*, we produce a set of *SEL*s whose locations correspond to these grid locations. However, if the grid connectivity is kept, and unit distances are used in the Laplacian, then convergence occurs. Figure 5.16 shows a close up of part of the grid, as well as the concentrations of morphogen a

Table 5-1. Mean Steps and Stripes.

number of cells	mean steps	mean stripes
10	720	1.35
20	883	3.14
30	938	4.83
40	955	6.62
50	998	8.30
60	1,041	9.89
70	1,047	11.54
80	1,086	13.16
90	1,102	14.80
100	1,113	16.49

Table 5-2. Mean Steps and Stripes.

max x/y offset	≤ 0.3	0.35	0.40	0.45	0.50	0.55	0.60	$0.65 \geq$
avg failures per trial	0.0	0.06	0.11	0.58	0.85	0.94	0.96	1.0

resulting from an execution of the reaction-diffusion code at each *SEL*. The concentration of morphogen a in the *S-Net* is displayed by producing an image array of specified size and assigning the appropriate gray level at each pixel according to the amount of morphogen a at each *SEL* located in the corresponding pixel. Also, note that in the *S-Net*, there is no wrap-around diffusion at the boundaries.

Similar to the 1D case, we find that if the locations of the *SEL*s are perturbed off the grid locations, and we use the actual inter-*SEL* distances to compute the reaction-diffusion equations, then the failure rate of convergence increases with distance from grid locations. Figure 5.17 shows a set of *SEL*s perturbed by up to 2 pixels in x and 2 pixels in y from the grid positions, but maintaining the same connectivity as the grid, and the corresponding pattern computed by the reaction-diffusion process. This is the pattern after 3,000 iterations, but it has not yet stabilized and the spots have not yet developed.

Next, we consider the case of *SEL*s randomly distributed in the square area. It turns out that if all neighbors within a certain distance (e.g., broadcast range) are used in the reaction-diffusion calculation, and the distances are used to compute the Laplacian, then the process generally fails to converge. However, if each *SEL* randomly selects four of its neighbors (e.g., from the broadcast connectivity graph), then the reaction-diffusion process converges. Figure 5.18 shows an example of this on a 5×5 square area using 2,000 *SEL*s placed by sampling the x and y coordinates from the uniform distribution. Figure 5.19 shows the layout of a set of 42 X-Bow T-Sky motes; with

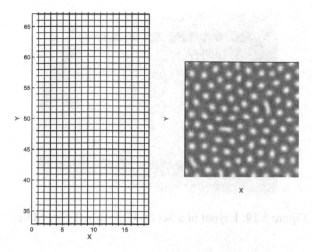

Figure 5.16: *S-Net* Regular Grid 2-D Turing Pattern.

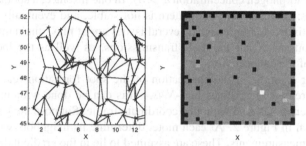

Figure 5.17: *S-Net* of Perturbed Regular Grid 2-D Turing Pattern.

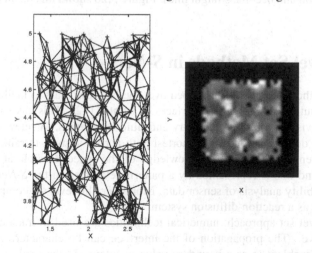

Figure 5.18: *S-Net* of Randomly Placed *SELs* and the Resulting 2-D Turing Pattern using 4 Randomly Selected Neighbors.

Figure 5.19: Layout of a Set of 42 X-Bow T-Sky Motes.

the small number of motes, the reaction-diffusion process leads to the formation of spots (i.e., the morphogen concentration $a > 5$). In one instance, a spot appeared after 1,250 iterations; however, the spot pattern is not stable, and eventually disappears if allowed to continue. We believe that several issues are at play: the number of nodes, the asynchronous nature of the mote transmissions, as well as the locality and bi-directionality of the connectivity.

One further example of the application of the spot formation reaction-diffusion process is to create straight lines in *S-Nets*. This can be done without knowing the distance between motes, and without a coördinate frame. Suppose the goal is to create stripes as shown in Figure 2. At each mote, determine its neighbors with lowest and highest sensor measurements. These are assumed to lie in the gradient direction. If the diffusion is restricted to take place only through these neighbors, then the spots blend in this direction and become straight lines. Figure 5.20 shows this result on a simulated set of motes.

5.3   Level Set Methods in *S-Net*s

Level set methods have been developed over the last few years to facilitate the analysis and computation of evolving interfaces or fronts [142]. This has wide application in computer vision, robotics, geometry and fluid mechanics. A relevant example for *S-Net*s is the determination of the shortest path across a given terrain. The level set methods depends in this case on knowledge of the speed possible at each point on the terrain, and this is represented by a pattern developed in the *S-Net*; e.g., by surface traversability analysis of sensor data. Moreover, the level set computation can be reformulated as a reaction-diffusion system.

In the level set approach, numerical techniques are used to track complex fronts as they evolve. The propagation of the interface can be characterized as either an initial value problem or as a boundary value problem. In the application of interest here, a curve in 2-D represents the front, and this curve moves outward with some speed and normal to the curve. The speed is either constant or a function of position.

Figure 5.20: Lines Produced by 2D Reaction-Diffusion Process.

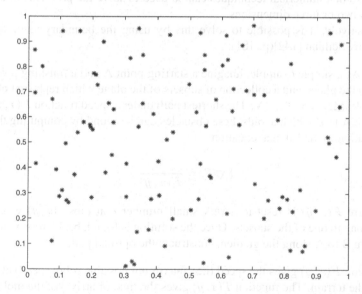

Figure 5.21: Shortest Path through Terrain Scenario (A:Start; B: Goal).

For example, a higher maximum speed is possible on a road than on rough terrain. Figure 5.21 shows the scenario: the *S-Net* is to compute a shortest path over a surface terrain. Each *SEL* produces an estimate of the terrain traversability index at its location. Based on these, as well as the start and goal positions, we seek to produce a time minimal path. Thus, the phenomenon model in this case is the motion of the front across the terrain. this means that we would like to determine optimal paths from the mobile robot to a target location with the following conditions:

- the computation should be distributed in the *SELs*

- the computation should be able to incorporate constraints (e.g., in the form of a speed function tied to the nature of the terrain)

- the computation should be robust.

In the initial value level set formulation, we assume there is a closed hyper-surface $\Gamma(t=0)$, propagating along its normal direction with speed F, which is a function of curvature, normal direction, etc. The main idea of the level set methodology is to embed this propagating interface as the zero level set of a higher dimensional function ϕ, defined as $\phi(x,t=0)$. x is a point in $\Re^n$. The equation for the evolving function $\phi(x,t)$ that contains the embedded motion of $\Gamma(t)$ as the zero level set (in the Eulerian formulation) is:

$$\phi_t + F|\nabla\phi| = 0 \tag{5.3}$$

This formula transforms a geometry problem into an initial value partial differential equation, so that numerical techniques can be used to solve the problems of a surface moving in two or three dimensions.

Alternatively, it is possible to solve this by using the boundary value approach proposed by Sethian [142](p. 181):

> As a simple example, imagine a starting point A and a finishing point B in the plane, and a collection of subsets of the plane which represent obstacles: $\Omega_j, j = 1 \dots N$. The shortest path under a speed function $F(x,y)$ from A to B which avoids these obstacles can be found by computing the solution to the Eikonal equation:
>
> $$\mid \nabla T \mid = \frac{1}{F(x,y)}$$
>
> where $F(x,y)$ is reset to a very small number ϵ at those (x,y) which belong to one of the subsets. Once the solution is found, back propagation from B to A along the gradient constructs the optimal path.

The function $F(x,y)$ gives the speed (i.e., maximum possible robot speed) at location (x,y) in the terrain. The function $T(x,y)$ gives the time of arrival of the mobile robot were it to take the fastest possible path to (x,y) from its current position.

In general, the function T depends on the mobile robot's location and must be computed. Once T is known, the gradient of T can be computed, and then the optimal path is found by following the opposite of the gradient of T from the target back to the robot. Our insight is that this computation is easily done by the distributed sensor devices.

To solve the shortest path problem, consider a boundary propagating through the plane, and let its time of arrival at a point (x,y) be $T(x,u)$; note the discussion here is based on Sethian's exposition). Level sets allow the determination of an optimal solution in terms of shortest arrival time. Assume that the maximum speed possible at every point is given by $F(x, y) > 0$. Since $d = rt$, we have:

$$dx = F(x, y)dT$$

$$\Rightarrow 1 = F(x, y)\frac{dT}{dx}$$

It is the case then that ∇T is orthogonal to the level sets of T (which are equal arrival times), and

$$|\nabla T| F = 1$$

Given that $T(x, y) = 0$ at the start location, then the boundary value formulation is:

$$front = \Gamma(t) = \{(x, y) \mid T(x, y) = t\}$$

When $F(x, y)$ depends only on position, then this is called the *Eikonal equation*.

5.3.1 Simple Level Set Example

Consider a terrain for which the robot speed is constant at every point in the plane, e.g., $F(x, y) = 1$. If the starting point is at the origin, then this represents a circular front and $T(x, y) = \sqrt{x^2 + y^2}$. Then the analytic solution is:

$$\nabla T(x, y) = \begin{bmatrix} \dfrac{\partial T}{\partial x} \\ \dfrac{\partial T}{\partial y} \end{bmatrix} = \begin{bmatrix} \dfrac{x}{\sqrt{x^2 + y^2}} \\ \dfrac{y}{\sqrt{x^2 + y^2}} \end{bmatrix}$$

The Eikonal equation is:

$$|\nabla T| = 1$$

and

$$\Gamma(t) = \{(x, y) \mid t = \sqrt{x^2 + y^2}\}$$

Thus, the level set is given by this implicit formula and for each t is the set of (x, y) points lying on the circle of radius t centered at the origin. Some of the advantages of this method include that it can be accurately approximated numerically, and that severe topological changes in the front (e.g., two circles crossing), are handled very well by the level set curve.

5.3.2 Shortest Path Problem

Let us now consider in more detail the shortest arrival time problem. Given:

- S: a set of n *SELs* with locations (x_i, y_i)
- F_i: the terrain speed value at *SEL* S_i
- A: start location (x_A, y_A)
- B: start location (x_B, y_B)

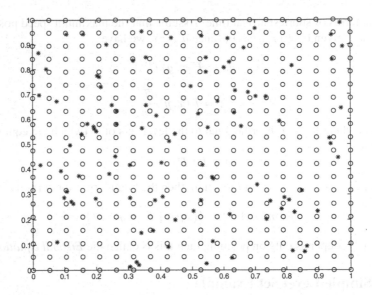

Figure 5.22: *SELs* with Overlayed Virtual Regular Grid.

Kimmel and Sethian [91] describe how to use level sets to find the shortest time of arrival path from A to B. The level set $T(x, y) = C$ is the set of all points in the plane which take minimal time C to reach. The Eikonal equation is solved to determine $T(x, y)$ at all points of interest. Subsequently, the shortest path is found by tracing back from B to A following the gradient.

The solution provided above is intended for a regularly spaced orthogonal grid aligned with the $x-$ and $y-$axes. Given the set of 100 *SELs* shown in Figure 5.21, we propose to construct a virtual regular grid of the desired spacing and to assign speed function values based on the neighboring *SELs*. Figure 5.22 shows an example 20x20 grid overlayed on the *SELs*. A simple first approach is to assign the speed of the nearest *SEL* to each grid point; this corresponds to the set of all grid points in the Voronoi cell of each *SEL*. Figure 5.23 shows this. Also shown are start point A at $(0.07, 0.93)$ and goal point B at $(0.92, 0.18)$.

Although the basic level set method requires expression of the gradient as a finite difference and iterative solution of the resulting quadratic equation at each point, we propose the following variant. Let G be the set of all grid points, V_i be the Voronoi region of each *SEL*, and G_i be the grid points in V_i.

Initialize

Let g be grid point closest to A

Set T(g) to 0 and add g to *FRONTIER*

Set T(h) = $infty$ for all other grid points h

Add all other grid points to *OPEN*

Set *CLOSED* to empty; set *Frontier* to $\{G - FRONTIER\}$

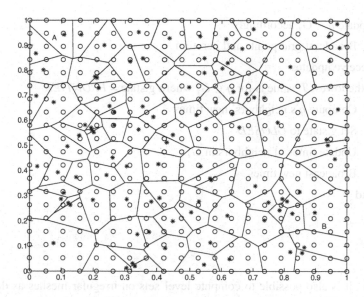

Figure 5.23: *SELs* with Grid Points and Voronoi Regions.

Loop until T(B) is determined

 Choose $N \in$ *FRONTIER* such that $T(N)$ is minimum

 Compute $T(M)$ for each neighbor M of N

 Replace $T(M)$ if new time is lower than previous

 If $M \in$ *OPEN*, move M to *FRONTIER*

 Move N to *CLOSED*

end loop

Figure 5.24 shows two example speed maps for the unit square, and the resulting short-est time of arrival path. The left map is constant speed and thus finds the straightest line from A to B. The right map has road-like areas and some obstacles (speed is zero in these regions).

 Of course, the shortest time of arrival algorithm must be modified some to run as the same algorithm in each *SEL*:

$\forall g \in G_i$, set g.state = *OPEN* and g.T = ∞

if $A \in V_i$, let g^* in G_i be closest to A

 Set g^*.state = *FRONTIER*, $g^*.T = 0$

Loop until $T(B)$ is determined

 if all $g \in G_i$ have g.state = *OPEN*

  Receive updates

Update $g \in G_i$ accordingly

else   Broadcast grid points and iteration

Receive updates

if the lowest T value is for $g \in G_i$ where g.state = *FRONTIER*

Compute g's neighbors' new times

Set g.state to *CLOSED*

Update times of neighbors in G_i

Broadcast new times

end

end

end loop

Of course, it is also possible to compute level sets on irregular meshes as described by Sethian [142], however, this requires even more sophisticated mathematical techniques, and their development in the *S-Net* is more difficult.

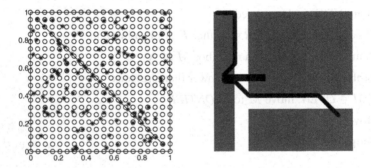

Figure 5.24: Shortest Paths for Two Different Speed Maps; (left: constant F; right: variable speeds: path goes to road (white) and down, then around obstacle blocking road, and then along road and then to goal).

5.4 Future Directions

We have described techniques for forming patterns in sensor networks based on trigonometric functions and reaction diffusion equations. Such methods may be used to process data, although that is not discussed here, and to carry sensed signal information. The next focus of our work is on the production of patterns based on sense data analysis (e.g., camouflage synthesis). Such methods may also find application in sensor network security; in this scenario, a deformed pattern will emerge from a distributed computation if there are any nodes which have fallen victim to attack, or if external nodes have managed to get themselves incorporated into the *S-Net*. Of course, resource allocation and exploitation may also be based on patterns, and given the random nature of the patterns, may help conserve resources (e.g., energy) overall.

We are also exploring mesh generation in wireless sensor networks. As was pointed out by Adamatzky et al. [1], physical reaction-diffusion computers can calculate Voronoi graphs. Thus, a basis exists in the reaction-diffusion computation to produce good triangulations for mote connectivity. The triangle are also important for computations on irregular meshes; e.g., for finite element methods. We are exploring this in the context of a larger set of motes.

Another area of research is the calculation of level sets [142] in the *S-Net*. These can be used for shortest path computation where an arbitrary speed function may be defined. We have shown how mobile robots can use this approach to find the lowest time path to traverse variable speed terrain [19]. However, the Eikonal equation used there may be set up as a reaction-diffusion system, and piggy-backed on the approach defined here.

Finally, stripes and spots may be of use for various purposes by mobile agents or the *S-Net*, but a more direct combination of the *S-Net* as both a sensing and display device is to be found in the creation of active camouflage. Consider the case of a camouflaged truck. Although it has a standard camouflage tarpaulin, it may not blend well with the forest behind it. Several problems may exist with coloration and blobs versus stripes (e.g., tree trunks and branches), leaf texture, etc.

One approach to overcome this mismatch is to pair a modeling process of the natural scene behind the truck with a display synthesis component in front of the truck. The technical basis for such a mechanism can be found in the work of Zhu and Mumford [170]. They propose (1) a theory for discovering the statistics of a set of natural images, and (2) a framework which allows the definition of reaction-diffusion equations to produce similar natural images, and in particular, they show how to remove conspicuously dissimilar segments from a scene. Specifically, they show that given a learned set of prior models that reproduce the observed statistics, the potentials of the resulting Gibbs distributions have potentials of the form:

$$U(I; \Lambda, S) = \sum_{\alpha=1}^{K} \sum_{x,y} \lambda^{\alpha}((F^{(\alpha)} \times I)(x, y))$$

where $S = \{F^{(1)}, F^{(2)}, \ldots, F^{(K)}\}$ is a set of filters and $\Lambda = \{\lambda^{(1)}, \lambda^{(2)}, \ldots, \lambda^{(K)}\}$ is the set of potential functions.

Reaction-diffusion equations are found as the gradient descent partial differential equations on $U(I; \Lambda, S)$; diffusion arises from the energy terms while pattern formation reactions are related to the inverted energy terms. These are then used to remove clutter in a scene and to denoise images. 5One example of this process given by Zhu We propose that the resulting images can be displayed by LEDs distributed throughout the material of the camouflage canvas, and based upon our previous work in e-textiles [83, 85, 110], we believe that technical solutions exist for the realization of this goal.

Chapter 6

Logical Sensors and Computational Mapping

Computational Sensor Networks[1] represent a scientific computing approach, and this includes the Verification and Validation (V & V) methodology of that discipline [118]; that is, model implementations must be verified (e.g., for numerical properties like error and convergence), and appropriate tests embedded in the system to monitor system correctness during execution. However, an important new aspect of this approach is that a CSN has the ability to sense and interact with the environment, and thus can run its own validation experiments to confirm or refute model structure or parameter values.

Given a computational method, such as one of those described previously, the next major goal is to develop a framework to facilitate mapping it onto a sensor network architecture. A strong requirement is to build on existing architectures (e.g., Sensor Network Architecture (SNA), Tenet, etc.) and to provide value added for information analysis methods (e.g., COMPASS and WaveScope). Our approach to this is to extend our previous Logical Sensor Specification (LSS) methodology into the CSN arena.

Recall that the SNA group has stated [24] that an architecture is not an end in itself, but that the task at hand is "the dense monitoring and analysis of sizable extents of the physical world," and that "sensor networks allow the understanding of the systematic behavior of various phenomena." The CSN approach allows a more quantified understanding of physical phenomena and the sensor network, too.

LSS corresponds to a *functional architecture* in terms of the SNA ([24], p. 2):

> *Functional Architecture*: This is the analog of "machine organization" in computer design or system decomposition in software systems. It includes a description of the logical building blocks or functional units, the capabilities of each, and their interconnection. In the TCP/IP stack, this is the suite of protocols and their interdependencies. While these Internet

---

[1]This chapter is a modified version of work done with Esther Shilcrat [65] (Section 6.1) and Mohamed Dekhil [31] (Section 6.2)

T.C. Henderson, *Computational Sensor Networks*, DOI: 10.1007/978-0-387-09643-8_6,
© Springer Science+Business Media, LLC 2009

protocols are largely arranged in a layered manner, it appears that the component services in sensor networks are more deeply interrelated, and the current set of designs have wide variations in both the decomposition and the interconnection.

Another relevant aspect of the SNA architecture to the LSS approach is the programming architecture which is composed of two aspects:

- *internal programming interface (IPI)*: how processing is performed across the numerous nodes of the sensor network, and

- *external programming interface (EPI)*: how information extracted from the sensor network is processed on conventional computing resources.

With respect to the IPI, CSN's require an organization of the task structure, as well as load balancing, communication minimization, etc. We have a great deal of experience with this in large-scale, multi-physics simulation [103], and we are developing a similar approach for CSN's.

Generally, the EPI is viewed as conventional, but in our case, certain aspects of CSN's require more detailed exchange of information through the interface: (1) error analysis, (2) physical phenomena models or computational methods, and (3) V & V requirements.

6.1 Logical Sensors

An early architectural approach which advocated strong programming semantics for multisensor systems is our Logical Sensor System (LSS) [65]. This approach exploits functional (or applicative) language theory to achieve that. The most developed version of LSS is Instrumented Logical Sensor Specification (ILSS) [31, 26]. We have developed Logical Sensors as specific sensor system specification methodologies whose overall goal is to aid in the coherent synthesis of efficient and reliable sensor systems.

Both the need for and availability of wireless sensor networks is growing, as is their complexity in terms of the number and variety of sensors within a system. Sensor networks now have a diversity of processors and sensors, and ad hoc techniques have been used to integrate them into a complete system and to operate on their data. In the future, however, such systems must be adaptable and reconfigurable to account for both sensor redundancy, complementarity and coverage, as well as to optimize time, space, and power efficiency. Two major issues regarding the configuration of *S Net*s arise:

1. how to develop a coherent and efficient treatment of the information provided by many *SEL*s, particularly when the sensors are of various kinds;

2. how to allow for *S-Net* reconfiguration, as a means toward greater tolerance for *SEL* device failure, dynamic selection of resources, and to facilitate future incorporation of additional *SEL*s.

The Multisensor Kernel System (MKS) was proposed as a uniform mechanism for dealing with data taken from several diverse sensors [54]. MKS has three major components: low-level data organization, high-level modeling, and logical sensor specification. The first two components of MKS concern the choice of a low-level representation of real-world phenomena and the integration of that representation into a meaningful interpretation of the real world, and have been discussed in detail elsewhere [38]. The logical sensor specification component aids the user in the configuration and integration of data such that, regardless of the number and kinds of sensor devices, the data are represented consistently. As such, the logical sensor specification component is designed in keeping with the overall goal of MKS, which is to provide an efficient and uniform mechanism for dealing with data taken from the *S-Net*, as well as facilitating sensor system reconfiguration. However, the logical sensor specification component of MKS can be used independently of the other two MKS components, and thus, a use for logical sensors is evident in any multisensor system and where sensor reconfiguration is desired.

The emergence of significant *S-Net* systems provides a major motivation for the development and application of logical sensors. Monitoring highly automated factories, complex chemical processes or the environment requires the integration and analysis of diverse types of sensor measurements; e.g., it may be necessary to monitor temperature, pressure, reaction rates, etc. In many cases, fault tolerance is of vital concern; e.g., in nuclear power plants, vehicles and other transportation structures. Our work has been done in a variety of contexts ranging from robotic workcells to snow quality monitoring, involving the following kinds of sensors:

- temperature sensors,
- cameras (an intensity array of the scene is produced),
- tactile pads (local forces are sensed),
- proximity sensors,
- laser range finders (distance to surfaces is produced),
- smart sensors (special algorithms are implemented in hardware).

Oftentimes if the special hardware is not available, then some of these sensors may be implemented as a software/hardware combination which should be viewed as a distinct sensor and which may ultimately be replaced by special hardware.

Other principal motivations for logical sensor specification are:

1. *Benefits of data abstraction*: the specification of a *SEL* is separated from its implementation. The *S-Net* is then much more portable in that the specifications remain the same over a wide range of implementations. Moreover, alternative mechanisms can be specified to produce the same sensor information but perhaps with different precision or at different rates. Further, the stress on modularity not only contributes to intellectual manageability [163], but is also an essential component of the system's reconfigurable nature. The inherent hierarchical structuring of logical sensors further aids system development.

2. *Availability of smart sensors*: the lowering cost of hardware combined with developing methodologies for the transformation from high-level algorithmic languages to silicon have made possible a system view in which hardware/software

divisions are transparent. It is now possible to incorporate fairly complex algorithms directly into hardware. Thus, the substitution of hardware for software (and vice versa) should be transparent above the implementation level.

Logical sensors are proposed as a means by which to insulate the user from the peculiarities of the sensor devices, which in this case are (generally) physical sensors. Thus, for example, a sensor system could be designed around camera input without regard to the kind of camera being used. However, in addition to providing insulation from the vagaries of physical devices, logical sensor specification is also a means to create and package *virtual* sensors. For example, the kind of data produced by a physical laser rangefinder sensor could also be produced by two cameras and a stereo program. This similarity of output result is more important to the user than the fact that one o getting it is by using one physical device, and the other way is by using two physical devices and a program. Logical sensor specification allows the user to ignore such differences of how output is produced, and treat different means of obtaining equivalent data as logically the same.

We have touched briefly on the rôle of logical sensors. We now define them formally. A *logical sensor* is defined in terms of four parts:

1. A *logical sensor name*. This is used to uniquely identify the logical sensor.

2. A *characteristic output vector*. This is basically a vector of types which serves as a description of the output vectors that will be produced by the logical sensor. Thus, the output of the logical sensor is a set (or stream) of vectors, each of which is of the type declared by that logical sensor's characteristic output vector. The type may be a standard type (e.g., real, integer, etc.), a user-generated type, or a well-defined subrange of either. When an output vector is of the type declared by a characteristic output vector (i.e., the cross product of the vector element types), we say that the output vector is an *instantiation* of the characteristic output vector.

3. A *selector* whose inputs are alternate subnets and an acceptance test name. The rôle of the selector is to detect failure of an alternate and switch to a different alternate. If switching cannot be done, the selector reports failure of the logical sensor.

4. *Alternate subnets*. This is a list of one or more alternate ways in which to obtain data with the same characteristic output vector. Hence, each alternate subnet is equivalent, with regard to type, to all other alternate subnets in the list, and can serve as backups in case of failure. Each alternate subnet in the list is composed of (a) a set of *input sources*, and (b) a *computation unit*. Each element of the set of input sources must itself be a logical sensor ot the empty set (null). Allowing null input permits *physical* sensors, which have only an associated program (the device driver), to be described as a logical sensor, thereby permitting uniformity of sensor treatment. A computation unit is a software program, or perhaps hardware units may also be used. In some cases a special *do nothing* computation unit may be used. We refer to this unit as **PASS**.

A logical sensor can be viewed as a network composed of subnetworks which are themselves logical sensors. Communication within a network is controlled via the flow of data from one subnetwork to another. Hence, such networks are *data flow* networks. Alternatively, we present the following inductive definition of a logical sensor. A logical sensor is an acceptance test which checks (sequentially and on demand) the output of either:

1. A list of computation units, with specified output type (the characteristic output vector), which require no input sources.

2. A list of computation units, with specified output type, whose input sources are logical sensors.

Figure 6.1 gives a pictorial presentation of this notion. The characteristic output vector declared for this logical sensor is (x-loc:real, y-loc:real, z-loc:real, curvature:integer). We present two examples to clarify the definition of logical sensors, and in particular to show how the inputs to a logical sensor are defined in terms of other logical sensors and how the program accepts input from the source logical sensors, performs some computation on them, and returns as output a set (stream) of vectors of the type defined by the characteristic output vector. Figure 6.2 shows the logical sensor specification for a *Camera* which happens to have no other logical sensor inputs. The specification for a stereo camera range finder called *Range Finder* is given in Figure 6.3. The program *stereo* takes the output of two cameras and computes vectors of the form (x, y, z) for every point on the surface of an object in the field of view. The idea is that a logical sensor can specify either a device driver program which needs no other logical sensor input and gets its input directly from the physical device, or a logical sensor can specify that the output of other logical sensors be routed to a certain program and the result packaged as indicated. This allows the user to create *packages* of methods which produce equivalent data, while ignoring the internal configurations of those packages.

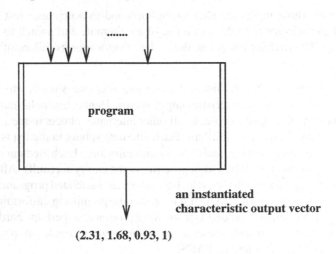

Figure 6.1: Graphical View of a Logical Sensor (adapted from [65]).

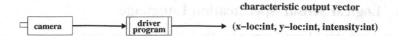

Figure 6.2: The Logical Sensor Specification of a Camera (adapted from [65]).

Figure 6.3: The Logical Sensor of a Range-Finder (adapted from [65]).

6.1.1 Formal Aspects of Logical Sensors

Having described how logical sensors are developed and operate, we now define a logical sensor to be a *network* composed of one or more subnetworks, where each subnetwork is a logical sensor. The computation units of the logical sensors are the nodes of the network. Currently, the network forms a rooted acyclic graph. The graph is rooted because, taken in its entirety, it forms a complete description of a single logical sensor (versus, for example, being a description of two logical sensors which share subnetworks). We also say that it is rooted because there exists a path between each subnetwork and a computation unit of the final logical sensor. Logical sensors may not be defined in terms of themselves; that is, no recursion is allowed; thus, the graph is acyclic.

All communication within a network is accomplished via the flow of data from one subnetwork to another. No other explicit control mechanism, such as the use of shared variables, alerts, interrupts, etc., is allowed. The use of such control mechanisms would decrease the modularity and independent operation of subnetworks. Hence, the networks described by the logical sensor specification language are data flow networks, and have the following properties:

- A network is composed of independently, and possibly concurrently, operating subnetworks.

- A network, or some of its subnetworks, may communicate with its environment via possibly infinite input or output streams.

- Subnetworks are modular.

Since the actual output produced by a subnetwork may depend on things like hardware failures (and because the output produced by the different subnets of a logical sensor are only required to have the same type), the subnetworks (and hence the network) are also indeterminate.

6.1.2   Logical Sensor Specification Language

We have shown that a logical sensor has the following properties:

- A logical sensor is a network composed of subnetworks which are themselves logical sensors

- A logical sensor may be defined only in terms of other, previously defined, logical sensors.

- A computation unit is an integral part of the definition of a logical sensor.

- A logical sensor produces output of the type declared by its characteristic output vector, and the declaration of the characteristic output vector is also an integral part of the definition of the logical sensor.

It should be noted that there may be alternate input paths to a particular sensor, and these correspond to the alternate subnets. But even though there may be more than one path through which a logical sensor produces data, the output will be of the type declared by the logical sensor's characteristic output vector.

With these points in mind, a language for describing the logical sensor system can be formed. We give the syntax below.

Syntax

(logical sensor)	→	(logical-sensor-name)
		(characteristic-output-vector)
		(selector)
		(alternate-subnet-list)
(logical-sensor-name)	→	(identifier)
(characteristic-output-vector)	→	(name-type-list)
(name-type-list)	→	(identifier):(type)
		{; (name-type-list) }
(selector)		(acceptance-test-name)
(alternate-subnet-list)	→	(computation-unit-name)(input-list)
		{(alternate-subnet-list)}
(acceptance-test-name)	→	(identifier)
(input-list)	→	(logical-sensor-list)|null
(logical-sensor-list)	→	(logical-sensor)
		{(logical-sensor-list)}
(computation-unit-name)	→	(identifier)

Semantics

Below we present the high-level description of the *operational* semantics (i.e., the execution effect) for each rule of the grammar.

1. A *logical sensor* declaration provides an associated name for the logical sensor used for identification purposes, and a characteristic output vector to declare the type of output for that logical sensor. A selector performs the test and switch after the acceptance test and the alternate subnet list establishes the alternative way of providing the characteristic output vector.

2. A *logical sensor name* declaration associates a (unique) identifier for the logical sensor.

3. A *characteristic output vector* establishes the type of output for the logical sensor.

4. A *name-type list* declaration establishes the precise nature of the output type as declared by the characteristic output vector. It consists of a cross product of types, with an associated name.

5. A *selector* declaration specifies the order in which the alternates in the alternate subnet list will be tested by the acceptance test.

6. An *alternate subnet list* declaration establishes a series of input sources, computation unit name tuples, thus making known which logical sensors and computation units are part of the definition of the logical sensor being declared.

7. An *output list* declaration establishes which legal input sources (either none or a series of logical sensors) are to be used as input to the computation unit.

8. A *logical sensor list* declaration establishes the set of logical sensors to be used as input.

9. A *computation unit name* declaration establishes the name of the actual program which will execute on the declared input sources.

10. An *acceptance test name* declaration establishes the name of the actual program which will be used to test the alternate subnets.

It is also possible to provide more formal semantics for the logical sensor specification language. Many works provide *denotational* semantics (i.e., semantic schemes which associate with each construct in the language an abstract mathematical object) for general data flow networks [80, 81, 89]. When such semantics have been given for the networks represented by logical sensors, we will be able to formally prove desired network properties; e.g., that the output of a specified logical sensor has particular properties of interest.

We have two implementations of the logical sensor specification language: a C version (called C-LSS) running under UNIX, and a functional language version (called FUN-LSS). These have been described elsewhere [65]. For example, FUN-LSS provides a logical sensor specification interface for the user and maintains a database of

S-expressions which represents the logical sensor definitions. The operations allowed on logical sensors include:

- *Create*: a new logical sensor can be specified by giving all the necessary information and is it inserted in the database.

- *Update*: an existing logical sensor may have certain fields changed; in particular, alternative subnets can be added or deleted, program names and the corresponding sensor lists can be changed.

- *Delete*: a logical sensor can be deleted so long as no other logical sensor depends on it.

- *Dependencies*: all logical sensor dependencies are shown.

6.1.3 Fault Tolerance

The Logical Sensor Specification Language has been designed in accordance with the view that languages should facilitate error determination and recovery. As we have explained, a logical sensor has a selector function which takes possibly many alternative subnets as input. The selector determines errors and attempts recovery via switching to another alternate subnet. Each alternate subnet is an input source – computation unit pair. Selectors can detect failures which arise from either an input source or the computation unit. Thus, the selector together with the alternate subnets constitute a failure and substitution device, that is, a fault tolerance mechanism, and both hardware and software fault tolerance is achieved. This is particularly desirable in light of the fact that "fault tolerance does not necessarily require diagnosing the cause of the fault or *even deciding whether it arises from the hardware or the software*" (emphasis added) [126]. In a multisensor system, particularly where continuous operation is expected, trying to determine and correct exact source of failure may be prohibitively time consuming.

Substitution choices may be based on either *replication* or *replacement*. Replication means that exact duplicates of the failed component have been specified as alternate subnets, In replacement, a different unit is substituted. Replacement of software modules has long been recognized as necessary for software fault tolerance, with the hope, as Randell states, that using a software module of independent design will facilitate coping "with the circumstances that caused the main component to fail" [126]. We feel that replacement of physical sensors should be exploited both with Randell's point in view and because extraneous considerations, such as cost, and spatial limitations as to placement ability are very likely to limit the number of purely backup physical sensors which can be involved in a sensor system.

Recovery Blocks

The recovery block is a means of implementing software fault tolerance [126]. A recovery block contains a series of alternates which are to be executed in the order listed. Thus, the first in the series of alternates is the *primary* alternate. An acceptance test is used to ensure that the output produced by an alternate is correct or acceptable. First the primary alternate is executed and its output scrutinized via the acceptance test.

If it passes, that block is exited, otherwise the next alternate is tried, and so on. If no alternate passes, control switches to a new recovery block if one on the same level or higher is available; otherwise, an error results.

Similarly, a selector tries, in turn, each alternate subnet in the list and tests each one's output via an acceptance test. However, while Randell's scheme requires the use of complicated recovery mechanisms (restoring the state, etc.), the use of a data flow model makes error recovery relatively easy. Furthermore, our user interface computes the dependency relation between logical sensors [143]. This permits the system to know which other sensors are possibly affected by the failure of a sensor.

The general difficulties relating to software acceptance tests, such as how to devise them, how to make them simpler than the software module being tested, and so on, remain. It is our view that some acceptance tests will have to be designed by the user, and that our goal is simply to accommodate the use of the test. Unlike Randell, we envision the recovery block as a means for both hardware and software fault tolerance, and hence we also allow the user to specify general hardware acceptance tests. Such tests may be base, or example, on data link control information, two-way handshaking, and other protocols. It is important to note that a selector must be specified even if there is only one subnet in a logical sensor's list of alternate subnets. Without at least the minimal acceptance test of a time out, a logical sensor could be placed on hold forever even when alternate ways to obtain the necessary data could have been executed. Given the minimal acceptance test, the selector will at least be able to signal failure to a higher-level selector which may then institute a recovery. However, we also wish to devise special schemes for acceptance tests when the basis for substitution is replacement. While users will often know which logical sensors are functionally equivalent, it is also likely that not all possible substitutions of logical sensors can be considered. Thus, we are interested in helping the user expand what is considered functionally equivalent. Such a tool could also be used to automatically generate logical sensors.

We give an example logical sensor network in Figure 6.4. This example shows how to obtain surface point data from possible alternate methods. The characteristic output vector of Range-Finder is $(x : real, y : real, z : real)$ and is produced by selecting one of the two alternate subnets and projecting the first three elements of the characteristic output vectors. The preferred subnet is composed of the logical sensor Image-Range. This logical sensor has two alternate subnets which both have the dummy computational unit **PASS**. **PASS** does not effect the type of the logical sensor. These alternatives will be selected in turn to produce the characteristic output vector $(x : real, y : real, z : real, i : int)$. If both alternates fail, the Image-Range sensor has failed. The Range-Finder then selects the second subnet to obtain the $(x : real, y : real, z : real)$ information from the Tactile-Range's characteristic output vector. If the Tactile-Range subsequently fails, then the Range-Finder fails. Each subnet uses this mechanism to provide fault tolerance.

6.1.4 Ramifications a Replacement Scheme

Many difficult issues arise when fault tolerance is based on a replacement scheme. Because the replacement scheme is instrumented through the use of alternate subnets, the user can be sure that the *type* of output will remain constant, regardless of the particular

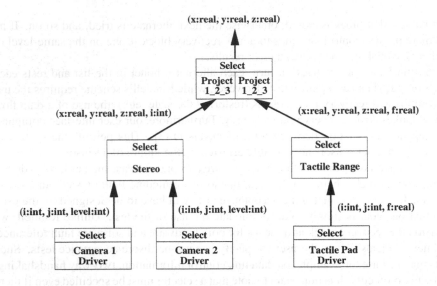

Figure 6.4: The Logical Sensor for Range-Finder (adapted from [65]).

source subnet. Ideally, however, we consider that a replacement-based scheme is truly fault tolerant only if the effect of the replacement is within allowable limits, where the allowable limits are determined by the user. As a simple example, consider a sensor array of one camera, A, and a backup camera, B, of another type. Suppose camera A has accuracy of ±0.01%, and camera B has accuracy of ±0.04%. If the user has determined that the allowable limit on accuracy is ±0.03%, then replacement of camera A by camera B will not yield what we call a truly fault tolerant system.

As mentioned above, determining functional equivalence may necessitate knowing more about a logical sensor than just its type. This example illustrates this point in that we have isolated a need to know about leaf logical sensors (physical sensors). However, we also mentioned that the above example was simplified. Let us now assume, in addition, that the user can use a variety of algorithms to obtain the desired final output. Suppose that one of those algorithms incorporates interpolation techniques which could increase the degree of accuracy of camera B's output. In this case, the user may use camera B and this algorithm as an alternate subnet and have a truly fault tolerant system. Thus, when we consider this more complex example, we see a general need for having features (besides type of output) of logical sensors visible, and a need to propagate such information through the system.

Feature propagation, together with allowable limit information, is needed for replacement-based fault tolerance schemes and constitutes an acceptance test mechanism. In addition, such feature propagation has a good potential for use in automatic logical sensor system specification and optimization. For example, consider an environment monitoring location with several *SEL*s. Once various logical sensors have been defined and stored, feature propagation can be used to configure new logical sensors with properties in specified ranges, or to determine the best logical sensor system (within the specified and perhaps weighted parameters).

6.1.5   Features and Their Propagation

Our view is that propagation of features will occur from the leaf nodes to the root of the network. In sensor systems, the leaf nodes will generally be physical sensors (with associated drivers). Thus, we first discuss the important features of physical sensors.

Features of Physical Sensors

Our goal here is to determine whether a set of generally applicable physical sensor features exists, and then to provide a database to support the propagation mechanism. In addition, it is possible for the user to extend the set of features. To date, existing systems provide a small set of generally applicable features.

All physical sensors convert physical properties or measurements to some alternative form, and hence are transducers. Some standard terms for use in considering transducer performance must be defined [164]. We have selected a set of features defined by Wright which we feel are generally applicable to physical sensors.

- *Error* – the difference between the value of a variable indicated by the instrument and the true value at the input.

- *Accuracy* – the relationship of the output to the true input within certain probability limits. Accuracy is a function of nonlinearities, hysteresis, temperature variation, etc.

- *Repeatability* – the closeness of agreement within a group of measurements at the same input conditions.

- *Drift* – the change in output that may occur despite constant input conditions.

- *Resolution* – the smallest change in input that will result in a significant change in transducer output.

- *Hysteresis* – a measure of the effect of history on the transducer.

- *Threshold* – the minimum change in input required to change the output from a zero indication.

- *Range* – the maximum range of input variable over which the transducer can operate.

Based on this set of physical sensor characteristics, the next step in arriving at a characterization of logical sensors is to compose physical sensor feature information with computation unit feature information.

Algorithm Features

There are several difficult issues involved in choosing a scheme whereby features of algorithms can be composed with features of physical sensors such that the overall logical sensor may be classified. As Bhanu [10] has pointed out: "The design of the system should be such that each of its components makes a maximum use of the input data characteristics and its goals are in conformity with the end result."

One issue to be resolved is how to represent features and feature composition. One approach is to record feature information and compositions functions separately. Thus, it would be necessary to classify an algorithm as having a certain degree of accuracy, and in addition, provide an accuracy function which given the accuracy of the physical sensor, produces the overall accuracy for the logical sensor which results from the composition of the physical sensor and the algorithm. A major difficulty in resolving such issues is presented by the great variety of sensor systems, both actual and potential, and the varying level of awareness of such issues within different sensor user communities. For example, experienced users of certain types of sensors may have a fairly tight knowledge of when and why certain algorithms work well, whereas other communities may not. Indeed, even within a sensor user community, algorithm evaluation techniques may not be standardized, hence yielding a plethora of ways in which properties of algorithms may be described. This problem is manifest in Bhanu's survey of the evaluation of automatic target recognition (ATR) algorithms.

The state of the art in algorithm evaluation techniques effects the choices made regarding the use of classifying physical sensors whether we wish to simply catalog information or maximize criteria. For example, if the user cannot provide information about the degree of resolution for the algorithms used, then an overall logical sensor resolution figure cannot be determined, even if the resolution of all physical sensors is known. Also, if such is the case, then the system cannot be used to help the user maximize the degree of resolution of the final output.

We now describe some techniques to allow for dynamic specification and allocation of logical sensors. Though the kinds of logical sensors which we consider represent only simple extensions to the existing logical sensor system, this type of work is the first step toward generally extensible logical sensor systems. The goal here is to show how, given information about logical sensors which can be configured in the system, new logical sensors can be defined automatically. Two techniques have been investigated: tupling and merging data.

Tupling Data

Tupling data is a technique which can be used to automatically generate new logical sensors in a feature-based sensor system. In such systems, the logical sensors return information about certain features found in the environment, such as objects present, motion, temperature, chemicals present, etc. The user may then request a new logical sensor be established by specifying the name for the new logical sensor, and giving the names of the input logical sensors. The output of the new logical sensor will be, simply, a set of tuples (one for each object in the environment), where each tuple is composed of the Cartesian product of the features which were input from the source logical sensors. Thus, we are basically packaging together features of interest so that they will be in one output stream. For example, suppose that the number of edges and number of holes are sufficient to determine the presence of bolts. Then a logical sensor *bolt detector* could be created by tupling the output of the logical sensors *edge detector* and *hole detector* (so long as they produce results which can be identified with specific objects). It should be noted that we assume that the latter two logical sensors produce output of the form (object number, feature 1, feature 2, ..., feature N). In this example, *hole detector* produces output of the form (object number, number of edges) and logical sensor *hole detector* produces output of the form (object number, number of holes). Logical sensor *bolt detector* will match the object number and produce tuples of the form (object number, number of edges, number of holes).

Merging Data

Another facility we have investigated dynamically incorporates, in response to a system demon, a newly defined logical sensor which outputs the merge of 3D logical sensor inputs. The idea is to accommodate an interactive request to allow the output of two physical sensors to be treated as one; for example, to create a multiple-view laser range finder logical sensor from two different laser range finder logical sensors. In this example, a logical sensor *multiview laser* is created with input logical sensors of both laser range finders, and the inputs are merged to produce output. Thus, the user can decide interactively to get more views without having to reconfigure the entire system. Also such a facility obviates the need for having multiple program units where the only difference is the number of expected inputs.

6.2   Instrumented Logical Sensor Systems

Instrumented Logical Sensor Systems (ILSS) were introduced as an extension to LSS which permits incorporation of the verification and validation aspects of computational science directly into the definition of the modules [29, 26]. Sensor systems are becoming ubiquitous throughout society, yet their design, construction and operation are still more of an art than a science. We define, develop, and apply a formal semantics for sensor systems that provides a theoretical framework for an integrated software architecture for modeling sensor-based control systems. Our goal is to develop a design framework which allows the user to model, analyze and experiment with different versions of a sensor system. This includes the ability to build and modify multisensor systems and to monitor and debug both the output of the system and the affect of any modification in terms of robustness, efficiency, and error measures. The notion of *Instrumented Logical Sensor Systems* (ILSS) that are derived from this modeling and design methodology is introduced. The instrumented sensor approach is based on a sensori-computational model which defines the components of the sensor system in terms of their functionality, accuracy, robustness and efficiency. This approach provides a uniform specification language to define sensor systems as a composition of smaller, predefined components. From a software engineering standpoint, this addresses the issues of modularity, reusability, and reliability for building complex systems. An example is given which compares vision and sonar techniques for the recovery of wall pose.

In any closed-loop control system, sensors are used to provide the feedback information that represents the current status of the system and the environmental uncertainties. Building a sensor system for a certain application is a process that includes the analysis of the system requirements, a model of the environment, the determination of system behavior under different conditions, and the selection of suitable sensors. The next step in building the sensor system is to assemble the hardware components and to develop the necessary software modules for data fusion and interpretation. Finally, the system is tested and the performance is analyzed. Once the system is built, it is difficult to monitor the different components of the system for the purpose of testing, debugging and analysis. It is also hard to evaluate the system in terms of time complexity, space complexity, robustness, and efficiency, since this requires quantitative measures for each of these.

In addition, designing and implementing real-time systems is becoming increasingly complex because of many added features such as fancy graphical user interfaces (GUIs), visualization capabilities and the use of many sensors of different types. Therefore, many software engineering issues such as reusability and the use of COTS (Commercial Off-The Shelf) components [125], real-time issues [71, 139, 145], sensor selection [44], reliability [84, 90, 150], and embedded testing [159] are now getting more attention from system developers.

We previously proposed to use formal semantics to define performance characteristics of sensor systems [27]. Here we address these and other problems related to sensor system modeling and evaluation. We start by presenting a theoretical framework for modeling and designing sensor systems based on a formal semantics in terms of a virtual sensing machine. This framework defines an explicit tie between the specification, robustness and efficiency of the sensor system by defining several quantitative measures that characterize certain aspects of the system's behavior. Figure 6.5 illustrates our proposed approach which provides static analysis (e.g., time/space complexity, error analysis) and dynamic handles that assist in monitoring and debugging the system.

6.2.1 Sensor Modeling

Each sensor type has different characteristics and functional description. Therefore it is desirable to find a general model for these different types that allows modeling sensor systems that are independent of the physical sensors used, and enables studying the performance and robustness of such systems. There have been many attempts to provide "the" general model along with its mathematical basis and description. Some of these modeling techniques concern error analysis and fault tolerance of multisensor systems [13, 32, 74, 112, 121, 122]. Other techniques are model-based and require a priori knowledge of the object and its environment [35, 48, 77]. These techniques help fit data to a model, but do not provide the means to compare alternatives. Task-directed sensing is another approach to devise sensing strategies [12, 51, 50], but again, it does not provide measures to evaluate the sensor system in terms of robustness and efficiency.

Another approach to modeling sensor systems is to define sensori-computational systems associated with each sensor to allow design, comparison, transformation, and reduction of any sensory system [34]. In this approach the concept of information invariants is used to define some measure of information complexity. This approach provides a very strong computational theory which allows comparing sensor systems, reducing one sensor system to another, and measuring the information complexity

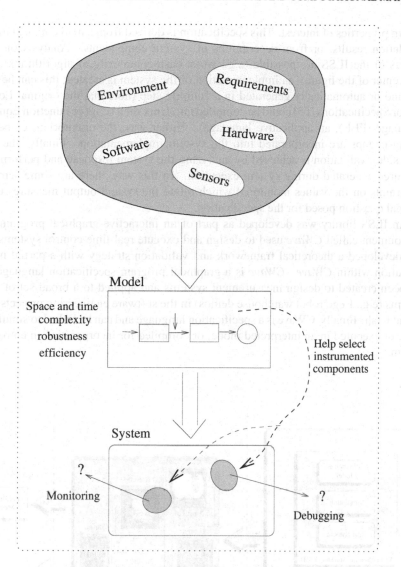

Figure 6.5: The Proposed Modeling Approach (adapted from [26]).

required to perform a certain task. However, as stated by Donald, the measures for information complexity are fundamentally different from performance measures. Also, this approach does not permit one to judge which system is "simpler," "better," or "cheaper."

To that end, we introduced the notion of an *Instrumented Logical Sensor System* (ILSS) which represents our methodology for incorporating design tools and allows static and dynamic performance analysis, on-line monitoring, and embedded testing. Figure 6.6 shows the components of our framework. First (on the left), an Instrumented Logical Sensor Specification is defined, as well as $\mathcal{F}$, a set of functions which measure system properties of interest. This specification is derived from a mathematical model, simulation results, or from descriptions of system components. Analysis of some aspects of the ILSS are possible (e.g., worst-case complexity of algorithms). Next (the center of the figure), an implementation of the system is created; this can be done by hand or automatically generated in a compile step (note that the original Logical Sensor Specifications [65] could be compiled into Unix shell script or Function Equation Language (FEL), an applicative language). Either way, the monitoring, embedded testing or taps are incorporated into the system implementation. Finally (the right hand side), validation is achieved by analyzing the system response and performance measures generated during system execution. In this way, there are some semantic constraints on the values monitored which relate the system output measures to the original question posed for the specification.

An ILSS library was developed as part of an interactive graphical programming environment called *CWave* used to design and execute real-time control systems. We also developed a theoretical framework and validation strategy with a partial implementation within *CWave*. *CWave* is a graphical program specification language that has been created to design measurement systems and applied to a broad set of robot systems (e.g., Lego robot warehouse demos) in the software engineering projects class here at Utah. Finally, CWave is a specification language and can be linked to simulation tools, or executed in an interpreted mode, or compiled for incorporation in embedded systems.

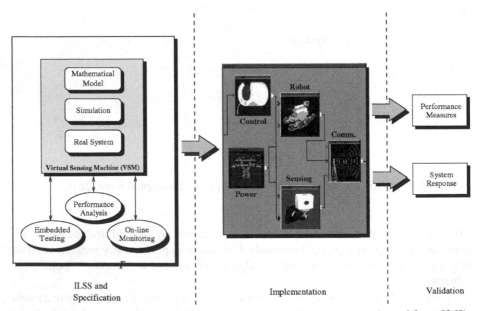

Figure 6.6: The Instrumented Logical Sensor System Components (adapted from [26]).

6.2.2   Performance Semantics of Sensor Systems

The use of sensors in safety critical applications, such as transportation and medicine, requires a high level of reliability. However, increased robustness and reliability of a multisensor system requires increased cost through redundant components and more sensor readings and computation. In contrast, increasing the efficiency of the system means less redundant components, fewer sensor readings and less computation. Performance analysis is crucial to making an informed tradeoff between design alternatives.

Performance analysis consists of a static analysis of a specification of the system and its parameters as well as a dynamic analysis of the system's run-time behavior. The static analysis can be based on some formal description of the syntax and semantics of the sensor system, while the dynamic analysis requires on-line monitoring of some quantitative measures during run-time.

Our goal is to achieve strong performance analysis and provide information which allows the user to make informed choices concerning system tradeoffs. This involves a sensor system model which permits quantitative measures of time and space complexity, error, robustness, and efficiency, and which facilitates analysis, debugging and on-line monitoring which directly supports the CSN paradigm.

Formal semantics of programming languages provides techniques to describe the meaning of a language based on precise mathematical principles. These formal techniques should provide the following: precise machine-independent concepts, unambiguous specification techniques, and a rigorous theory to support reliable reasoning [45]. The main types of formal semantics are: *denotational semantics* which concerns designing denotations for constructs, *operational semantics* which concerns the specification of an abstract machine together with the machine behavior when running the program, and *axiomatic semantics* which concerns axioms and rules of inference for reasoning about programs.

Our view is that performance semantics should allow us to compute measures of interest on program structures. Denotational semantics is the closest to our view since, according to [3], to specify the semantics of a language denotationally means to specify a group of functions which assigns mathematical objects to the program and to parts of programs (modules) in such a way that the semantics of a module depends only on the semantics of the submodules. Thus, given a set of programs, $\mathcal{P}$, from a language, and an operating context, $\mathcal{C}$, the semantics is a set of functions

$$\mathcal{F} = \{f_i\}$$

where

$$f_i : \mathcal{P} \times \mathcal{C} \to \Re$$

where $\Re$ is the measurement domain.

The static semantics defines structural measures over the syntax of $p \in \mathcal{P}$. This includes standard measures such as maximum depth of the program graph, branching measures, data structure properties, storage estimates and standard computational complexity measures. Note that these can be determined without reference to $\mathcal{C}$ (i.e., $f : \mathcal{P} \to \Re$). This can be extended to include functions of the operational context $\mathcal{C}$, including sensor models, accuracy, precision, redundancy and replacement, as well as operating system effects, communication strategies and protocols, and processor properties.

The dynamic semantics include validity measures and operational characteristics. Validity measures permit the comparison of behavior models to actual run-time performance (monitors), while operational characteristics are simply measures of run-time values (taps). The values of a tap or monitor are represented as a sequence $X = (x_n : n \in \mathcal{N})$; x_n is the nth value produced by the tap or monitor

$$X : \mathcal{N} \to S$$

where S is the structure produced by the tap or monitor.

The selection of functions in $\mathcal{F}$ depends directly on the user's needs and are defined so as to answer specific questions. Standard questions include actual running times, space requirements, bottlenecks, etc., and a complex application can be investigated in a top down manner – the user may define new measurement functions on lower level modules once information is gained at a higher level. This forces the user to identify crucial parameters and to measure their impact. For example, a target tracking application may be data dependent, say on the number of segmented objects or their distribution in the scene. Thus, the user is coerced into a better understanding of the significant value regimes of these parameters and may develop monitors to ensure that the application stays within a given range, or that it dynamically switches algorithms when a particular parameter value occurs (e.g., more than 100 segmented objects occur in the image). The main point is that the user can construct executable versions of the $f_i \in \mathcal{F}$ to ensure the validity of the controller as it runs.

Although computational complexity provides insight for worst case analysis, and for appropriate population distribution models, average case analysis can be performed, and we propose here what might be termed *empirical case analysis* which allows the user to gain insight into the system without requiring a detailed analytical model of the entire application and its context. Very few users exploit formal complexity analysis methods; we believe that empirical case analysis is a very useful tool.

Simple Example: Time Vs. Robustness Using Sonar Readings

Suppose that we have two mobile *SELs* and want to determine how many sonar readings to use to get a robust range estimate, but would like to trade off against the time taken to sample. This simple example demonstrates the motivation of the proposed approach and how it can be used to select between alternatives. In this example we have a "classical" tradeoff between speed (time to accomplish a certain task) and robustness (a combination of accuracy and repeatability). Assume that the sonar has been calibrated to eliminate any environmental effects (e.g., wall type, audio noises, etc.). The variables in this case are the accuracy of the physical sonar sensor and the number of readings taken for the same position.

Assuming the time to take one reading is t, the error standard deviation is σ, and the probability of a bad reading is Pr_b, taking one reading yields minimum time and worst accuracy. By adding a filter (e.g., averaging) and taking multiple readings, accuracy increases and time also increases. Therefore, we need quantitative measures to decide how many readings are needed to achieve the required accuracy (measured in terms of the standard deviation of the error) within a time limit.

Using the formalism presented earlier, the semantics of this problem can be defined using the set of functions $\mathcal{F} = \{time, error, repeatability\}$. In the case of using a single reading these functions can be written as:

$$time(single) = t$$

$$error(single) = \frac{\sigma}{\sqrt{(1 - Pr_b)}}$$

$$repeatability(single) = 1 - Pr_b$$

Now, if we take the average of n readings, the semantics can be written as:

$$time(average) = nt + \tau_n$$

$$error(average) = \frac{\sigma}{\sqrt{n * (1 - Pr_b)}}$$

$$repeatability(average) = 1 - Pr_b^n$$

where τ_n is the time to calculate the average of n readings, and $\tau_1 = 0$.

In this simple example we were able to get estimates of the required measures using mathematical models. However, we did not consider the changes in the environment and how it affects these measures. In this case, the set of functions $\mathcal{F}$ are mappings from the cross product of the program $\mathcal{P}$ and the operating context C to the measurement domain $\Re$, that is

$$f_i : \mathcal{P} \times C \to \Re$$

To solve this problem, we either have to model the environmental effects and include it in our model, or we may need to conduct simulations if a mathematical model is not possible. Simulation is a very useful tool to approximate reality, however, in some cases even simulation is not enough to capture all the variables in the model, and real experiments with statistical analysis may be required to get more accurate results. Thus, the formal functions can be operationalized as monitors or taps in the actual system.

6.3   Sensor System Specification

The ILSS approach is based on *Logical Sensor Systems* (LSS) described in the previous section, and is comprised of the following components (see Figure 6.7):

1. *ILS Name*: uniquely identifies a module.

2. *Characteristic Output Vector (COV)*: strongly typed output structure. We have one output vector (COV_{out}) and zero or more input vectors (COV_{in}).

3. *Commands*: input commands to the module ($Commands_{in}$) and output commands to other modules ($Commands_{out}$).

4. *Select Function*: selector which detects the failure of an alternate and switches to another alternate (if possible).

5. *Alternate Subnets*: alternative ways of producing the COV_{out}. It is these implementations of one or more algorithms that carry the main functions of the module.

6. *Control Command Interpreter (CCI)*: interpreter of the commands to the module.

7. *Embedded Tests*: self testing routines which increase robustness and facilitate debugging.

8. *Monitors*: modules that check the validity of the resulting COVs.

9. *Taps*: hooks on the output lines to view different COV values.

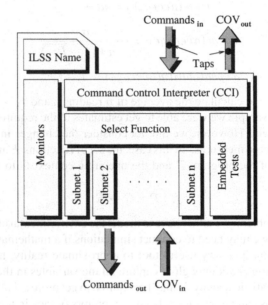

Figure 6.7: The Extended Logical Sensor Module (adapted from [26]).

These components identify the system behavior and provide mechanisms for on-line monitoring and debugging. In addition, they give handles for measuring the run-time performance of the system.

Monitors are validity check stations that filter the output and alert the user to any undesired results. Each monitor is equipped with a set of rules (or constraints) that governs the behavior of the COV under different situations.

Embedded testing is used for on-line checking and debugging proposes. Weller proposed a sensor processing model with the ability to detect measurement errors and to recover from these errors [159]. This method is based on providing each system module with verification tests to verify certain characteristics in the measured data and to verify the internal and output data resulting from the sensor module algorithm. The

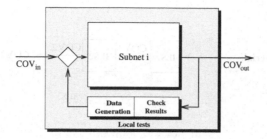

Figure 6.8: Local Embedded Testing (adapted from [26]).

recovery strategy is based on rules that are local to the different sensor modules. We use a similar approach in our framework called *local embedded testing* in which each module is equipped with a set of tests based on the semantic definition of that module. These tests generate input data to check different aspects of the module, then examine the output of the module using a set of constraints and rules defined by the semantics. Also these tests can take input data from other modules if we want to check the operation for a group of modules.

Figure 6.8 illustrates the idea of local embedded testing. Local embedded testing increases the robustness of the system and provides the user with possible locations to tap into when there is a problem with the system.

6.3.1   Construction Operators

In our proposed framework, a sensor system is composed of several ILSS modules connected together in a certain structure. We define operations for composing ILSS modules, and then define the semantics of these operations in terms of the performance parameters. Some of these operations are (see Figure 6.9):

- $Serial(ILSS1, ILSS2)$: two logical modules are connected in series. Here $COV3 = COV2$.

- $Select(ILSS1, ILSS2)$: $COV3$ is equal to either $COV1$ or $COV2$.

- $Combine(ILSS1, ILSS2)$: $COV3$ is the concatenation of $COV1$ and $COV2$.

For these simple constructs, the semantics is defined as a set of functions that propagate the required performance measures. Several techniques can be used for propagation. Best case analysis, worst case analysis, average, etc. Selecting among these depends on the application, hence it should be user defined. As an example, the time of the resulting logical system using worst case analysis can be calculated as follows:

- $time(Serial(ILSS1, ILSS2)) = time(ILSS1) + time(ILSS2)$

- $time(Select(ILSS1, ILSS2) = max(time(ILSS1), time(ILSS2))$

- $time(Combine(ILSS1, ILSS2) = max(time(ILSS1), time(ILSS2))$

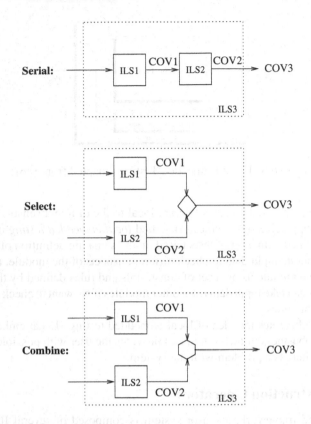

Figure 6.9: Some Operations used for Propagating the Performance Measures (adapted from [26]).

Hence, the semantic functions of the composite system are defined in terms of the semantic functions of the subcomponents, Similarly, functions that define the propagation of other performance measures can be defined in the same way.

For error propagation, we use a simple approach which does not require carrying a lot of information through the system. This approach is based on the uncertainty propagation described in [39, 70]. Assume that we have a certain module with n inputs $X = (x_1, x_2, \ldots, x_n)$ and m outputs $Y = (y_1, y_2, \ldots, y_m)$ such that $Y = f(X)$, and assume that the error variance associated with the input vector is $\Lambda_X = (\Lambda_{x_1}, \Lambda_{x_2}, \ldots, \Lambda_{x_n})$ (see Figure 6.10), then the error variance for the output vector is calculated using the equation:

$$\Lambda_Y = \left(\frac{\partial Y}{\partial X} \right) \Lambda_X \left(\frac{\partial Y}{\partial X} \right)^T$$

where $\frac{\partial Y}{\partial X}$ is the partial derivative of Y with respect to X evaluated at the measured value of the input vector X. If all the elements in X are independent variables, then this equation can be written as:

$$\Lambda_{y_i} = \sum_{j=1}^{n} \left(\frac{\partial y_i}{\partial x_j} \right)^2 \Lambda_{xj} \,, i = 1, 2, \ldots, m$$

Our overall goal is to provide a tightly coupled mechanism to map high-level performance measures onto an appropriate set of monitors, tests and taps so as to provide the required information.

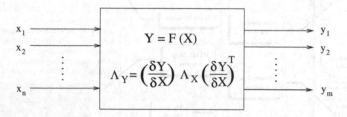

Figure 6.10: A Simple Approach for Error Propagation (adapted from [26]).

6.3.2 Implementation

The ultimate goal of this project is to utilize the proposed theoretical framework in a usable modeling and prototyping environment with tools for analysis, debugging, and monitoring sensor systems with emphasis on robot control applications. Thus, we developed an ILSS library within a visual programming system called CWave targeted toward the development of control systems for measurement devices and hardware simulations. CWave is developed by the Component Software Project (CSP) research group in the Department of Computer Science at the University of Utah in coöperation with the CSP group at Hewlett Packard Research Labs in Palo Alto, California.

CWave is based on a reusable software components methodology where any system can be implemented by visually wiring together predefined and/or user created components and defining the dataflow between these components. The CWave design environment includes several important features that make it suitable to use as a framework for implementing ILSS components. Some of these features are:

- Open architecture with ease of extensibility.

- Drag-and-drop interface for selecting components.

- Several execution modes including single step, slow, and fast execution.

- On-line modification of component properties.

- The ability to add code interactively using one of several scripting languages including Visual Basic and Java Script. This is particularly useful to add monitors and/or taps on the fly.

- Parallel execution using visual threads.

- On-line context sensitive help.

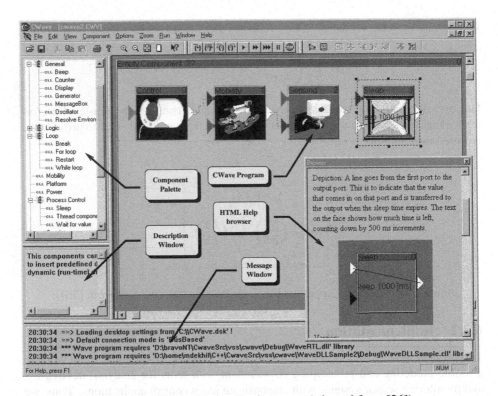

Figure 6.11: CWave Design Environment (adapted from [26]).

Figure 6.11 shows the CWave design environment with some of its features.

An object-oriented approach is used to develop the ILSS components using Visual C++ for implementation. Each component is an object that possesses some basic features common to all components plus some additional features that are specific to each ILSS type. The following are some of the basic functions supported by all components:

Initialize: performs some initialization steps when the component is created.

Calibrate: starts a calibration routine.

Sense: generates the COV corresponding to the current input and the component status.

Reset: resets all the dynamic parameters of the component to their initial state.

Test: performs one or more of the component's embedded tests.

Select: selects one of the alternate subnets. This allows for dynamic reconfiguration of the system.

Monitor: observes the COV and validate its behavior against some predefined characteristic criteria.

Tap: displays the value of the required variables.

We used several design patterns in designing and implementing the components. Design patterns provide reliable and flexible object-oriented designs that can accommodate rapid modifications and extensions [42]. For example, the *decorator* pattern is used to dynamically attach additional functionality to the object. This is particularly useful in our case where the user can dynamically choose the performance measures to be propagated and the values to be monitored while the system is running. Note that monitors, tests, and taps can be exploited to analyze CWave (or any implementation language) module performance independently of the sensor aspects of the system. This is rendered more efficient and transparent to the user by incorporating them directly as language features.

6.4   Example: Wall Pose Estimation

The following example illustrates the use of the proposed framework to model and analyze two alternatives for determining flat wall position and orientation: one using vision and one using sonar sensors [28, 59, 61, 30]. The sonar sensors are mobile *SEL*s (the experiments were carried out on a LABMATE mobile robot designed by Transitions Research Corporation).

In this example, we consider two different logical sensors to determine wall pose and find the corresponding errors and time complexity for each. The first ILSS uses a camera and known target size and location. The second ILSS deals with the sonar sensor as a wedge sensor (i.e., it returns a wedge centered at the sonar sensor and spread by an angle 2θ.) Figure 6.12 shows the two logical sensors. (See [61] for an overview of the sonar pose recovery technique, and [60] for target-based calibration.)

In this figure, *image* is the 128x128 black and white image acquired by the *Camera*, and r_1 and r_2 are the two sonar readings generated from *Sonar*1 and *Sonar*2, respectively. *Target Points* extracts three reference points from the *image*, while *Vision Line* produces two points on the line of intersection of the wall with the x-z plane of the camera system. $Wedge\_Sonar\_Line$ takes the two range values r_1 and r_2, and the spread angle of the sonar beam θ, and returns two 2D points on the line representing the wall.

6.4.1   System Modeling and Specification

As shown in Figure 6.12, ILSS1 is composed of three modules, a *Camera* module, a *Target Points* module and a *Vision Line* module. On the other hand, LSS2 has three modules, two *Sonar* modules and a $Wedge\_Sonar\_Line$ module followed by a $Combine$ operator.

Each ILSS is defined in terms of a set of components that characterize the module. The data and the corresponding performance measures start from the $Camera$ or $Sonar$ module and propagate upward until they reach the COV of the main ILSS. On the other hand, the commands start from the main ILSS and propagate downward until they reach the $Camera$ or $Sonar$ module. The COV is composed of two parts: *data* and *performance measures*. For example, COV_{out} for $Sonar$1 is

$$(\{r_1, \theta\}, \{t, \Lambda_{r1}, \Lambda_\theta\})$$

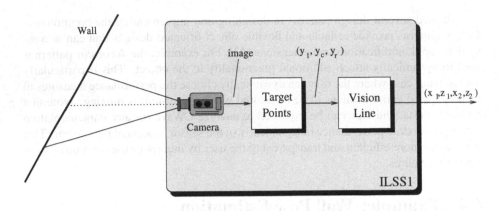

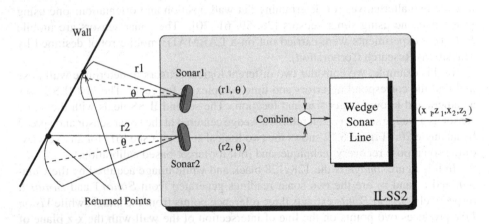

Figure 6.12: Two Instrumented Logical Sensors for Determining Wall Position (adapted from [26]).

where t is the time taken to execute the module and Λ_{r1} and Λ_θ are the error variances for r_1 and θ, respectively. In this example, each module has only one alternate subnet, therefore, the select function is trivial.

6.4.2 Performance Semantic Equations

Using worst case analysis, the performance semantic equations of the *time* and *error* for ILSS1 and ILSS2 can be written as:

$$time(ILSS1) = time(Serial(Camera, TargetPoints, VisionLine))$$

$$error(ILSS1) = error(Serial(Camera, TargetPoints, VisionLine))$$

$$time(ILSS2) = time(serial(combine(Sonar1, Sonar2), Wedge\_sonar\_line))$$

$$error(ILSS2) = error(serial(combine(Sonar1, Sonar2), Wedge\_sonar\_line))$$

Now, we need to calculate the time and error for the subcomponents. Assume that t_{sonar1}, t_{sonar2}, t_{camera}, $t_{TargetPoints}$, $t_{VisionLine}$ and $t_{wedge\_sonar\_line}$ are the time for the subcomponents, and Λ_{r1}, Λ_{r2}, Λ_{y_l}, Λ_{y_c}, Λ_{y_r} and Λ_θ are the error measures for r_1, r_2, y_l, y_c, y_r and θ, respectively. The time for LSS1 and LSS2 can be easily calculated using the propagation operations discussed earlier as follows:

$$time(ILSS1) = t_{camera} + t_{TargetPoints} + t_{VisionLine}$$

$$time(ILSS2) = max(t_{sonar1}, t_{sonar2}) + t_{wedge\_sonar\_line}$$

Propagating the error requires more elaborate analysis for each component. For ILSS1, we start with the error in the physical sensor which is the camera in this case. The camera generates two-dimensional arrays of intensity values, $P(x, y)$, where P is an $m \times n$ matrix. The error we are concerned abound in this example is the error in position (x, y) of a point on the CCD array (which corresponds to rows and columns in the image.) This error is affected by the resolution of the camera and the distance between the CCD elements. Let's assume that the error is Gaussian with mean 0 and variance (Λ_x, Λ_y) at any point (x, y). This can be written as:

$$error(Camera) = \{(\Lambda_x, \Lambda_y)_{m \times n}\}$$

This error translates directly into the second component, $Target\_Points$, which extracts the y value for three different points in the image; y_l, y_c, and y_r. Assuming that the variance in the y direction (Λ_y) is the same at any pixel, the error at this stage will be:

$$error(Target\_Points) = \{\Lambda_y, \Lambda_y, \Lambda_y\}$$

The last component in ILSS1, $Vision\_Line$ performs several operations on these three values to generate the two points of the line representing the wall. First, the corresponding z value is calculated for the three points using the equation:

$$z_i = \frac{Y_0}{y_i}, \qquad i = l, c, r$$

where Y_0 is the height of the physical point and is a known constant in our example. The error associated with z_i can be calculated as follows:

$$\Lambda_{z_i} = \left(\frac{\partial z_i}{\partial y_i}\right)^2 \Lambda_{y_i}$$

By calculating the derivative in the above equation we get:

$$\Lambda_{z_i} = \left(\frac{-Y_0}{y_i^2}\right)^2 \Lambda_y = \frac{Y_0^2}{y_i^4} \Lambda_y$$

which shows how Λ_{z_i} depends on the value of y_i. Second, the angle between the robot and the wall (α) is calculated with the function:

$$\alpha = sin^{-1}\left(\frac{z_l - z_r}{D_0}\right)$$

where D_0 is the known distance between the two physical points p_l and p_r. Therefore,

$$\Lambda_\alpha = \left(\frac{\partial \alpha}{\partial z_l}\right)^2 \Lambda_{z_l} + \left(\frac{\partial \alpha}{\partial z_r}\right)^2 \Lambda_{z_r}$$

$$= \left(\frac{1}{\sqrt{1 - \left(\frac{z_l - z_r}{D_0}\right)^2}}\right)^2 \Lambda_{z_l} + \left(\frac{-1}{\sqrt{1 - \left(\frac{z_l - z_r}{D_0}\right)^2}}\right)^2 \Lambda_{z_r}$$

After simplifying the last equation we get:

$$\Lambda_\alpha = \frac{D_0^2}{D_0^2 - (z_l - z_r)^2}(\Lambda_{z_l} + \Lambda_{z_r})$$

Finally, we calculate two points on the line representing the wall as shown in Figure 6.13. Take the first point p_1 at $(0, z_c)$ and the second point p_2 at one unit distance from p_1 along the wall which gives the point $(\cos \alpha, z_c + \sin \alpha)$:

$$x_1 = 0, \qquad z_1 = z_c$$

$$x_2 = \cos \alpha, \qquad z_2 = z_c + \sin \alpha$$

From these equations, the error for the two points will be:

$$\Lambda_{x_1} = 0, \qquad \Lambda_{z_1} = \Lambda_{z_c}$$

$$\Lambda_{x_2} = sin^2\alpha\, \Lambda_\alpha, \qquad \Lambda_{z_2} = \Lambda_{z_c} + cos^2\alpha\, \Lambda_\alpha$$

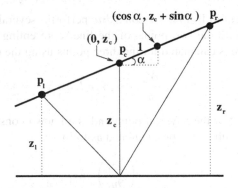

Figure 6.13: The Two Points on the Line Representing the Wall (adapted from [26]).

Now, we can write the error of ILSS1 as:

$$error(ILSS1) = \{\Lambda_{x_1}, \Lambda_{z_1}, \Lambda_{x_2}, \Lambda_{z_2}\}$$

Notice that we can write the error in terms of $\Lambda_y, Y_0, D_0, y_l, y_c$, and y_r. For example, let's assume that $\Lambda_y = 1mm^2, Y_0 = 500mm, D_0 = 300mm$, and $y_l = y_c = y_r = 10mm$ (α is zero in this case), then the error will be:

$$error(ILSS1) = \{0, 25mm^2, 0, 25mm^2\}$$

Now we analyze ILSS2 in a similar manner. At the first level, we have the physical sonar sensor where the error can be determined either from the manufacturer specs, or from experimental data. In this example we will use the error analysis done by Schenkat and Veigel [137] in which there is a Gaussian error with mean μ and variance σ^2. From this analysis, the variance is a function of the returned distance r. To simplify the problem let's assume that the variance in both sensors is $\Lambda_r = 4.0mm^2$. Therefore we can write the error in the sonars as:

$$error(Sonar) = \{\Lambda_r\}$$

In the $Wedge\_Sonar\_Line$ module, there are five possible cases for that line depending on the values of r_1 and r_2 [61]. In any case, the two points on the line can be written as:

$$x_1 = r_1 \cos \alpha_1, \qquad z_1 = r_1 \sin \alpha_1$$

$$x_2 = r_2 \cos \alpha_2, \qquad z_2 = r_2 \sin \alpha_2$$

where the values of $\alpha 1$ and α_2 are between $-\theta$ to θ (see Figure 6.14).

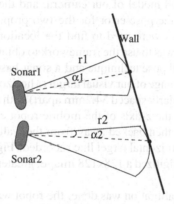

Figure 6.14: The General Case for the Points Returned by the Wedge_sonar_line (adapted from [26]).

Considering the worst case error, we can set $\alpha_1 = \alpha_2 = \theta$. Assuming that the error in θ is zero, then the error in the calculated points is:

$$\Lambda_{x_i} = \left(\frac{\partial x_i}{\partial r}\right)^2 \Lambda_r$$

$$\Lambda_{z_i} = \left(\frac{\partial z_i}{\partial r}\right)^2 \Lambda_r$$

which results in:

$$\Lambda_{x_1} = \cos^2 \theta \, \Lambda_r, \qquad \Lambda_{z_1} = \sin^2 \theta \, \Lambda_r$$

$$\Lambda_{x_2} = \cos^2 \theta \, \Lambda_r, \qquad \Lambda_{z_2} = \sin^2 \theta \, \Lambda_r$$

Finally, the error function for $ILSS2$ is:

$$error(ILSS2) = \{\Lambda_{x_1}, \Lambda_{z_1}, \Lambda_{x_2}, \Lambda_{z_2}\}$$

As an example, if $\Lambda_r = 4.0mm^2$, and $\theta = 11^o$ (approximately correct for the Polaroid sensor), we get:

$$error(ILSS2) = \{3.85mm^2, 0.15mm^2, 3.85mm^2, 0.15mm^2\}$$

This example illustrates the importance and usefulness of the ILSS library since all these analyses can be performed once and put in the library for reuse and the user does not have to go through these details again. For example, if a different sonar sensor is used, then the same error analysis can be used by supplying the sensor's error variance. In addition, given that the error range has been determined, redundancy can be added using different sensor pairs to sense the same wall and a monitor can be added to detect error discrepancies.

6.4.3 Experimental Results

We do not have a very good model of our camera, and therefore actual experiments were required to compare the pose error for the two proposed techniques. The two instrumented logical sensors were used to find the location of walls using real data. The goal of the experiment was to use the framework to obtain measures to help choose between a vision based wall pose technique and a sonar based wall pose estimator.

First, we calibrated the range of our visual target (a horizontal line at a known height, Y_0 with vertical stripes regularly spaced 34.2mm apart) with its y-location in the image. This was done by aligning the z-axis of the mobile robot camera to be normal to the wall; the mobile robot was then backed away from the wall a known distance and the image row number of the horizontal target line recorded. Figure 6.15 shows the results of this step. (Note that we digitized a 128x128 image; greater resolution would produce more accurate results.)

Once the target range calibration was done, the robot was placed in eight different poses with respect to the wall and the visual target acquired. Each image was constrained to have at least two vertical stripes and neither of them could be centered on the middle column of the image. The test images are shown in Figure 6.16.

Sonar data was also taken at each pose. The actual pose of the mobile robot with respect to the wall was independently measured by hand. Table 6.1 gives the hand measured, sonar and image calculated results. The error values of the sonar and vision results with respect to the handmeasured data are plotted in Figures 6.17 and 6.18.

These results allow the user to decide whether to use one technique or the other given the global context. For example, our application was a tennis ball pickup competition in which we were using vision to track tennis balls anyway, and we needed to locate a delivery location along the wall; if we can get by with pose error of less than $0.3m$ range and 15° angle, then ILSS1 will suffice. If less error were required, then a costly sonar system with hardware and software would need to be added to the robot, or else the use of higher resolution imagery could be explored. However, decisions made with respect to all these considerations would now be defensible and well documented.

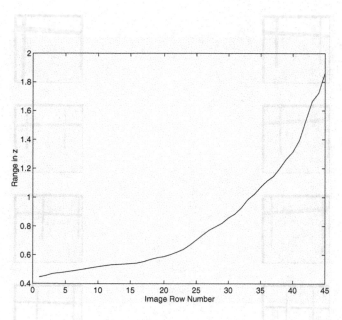

Figure 6.15: Row vs. Range (adapted from [26]).

Table 6.1: Pose Results from Measured Data, Sonar, and Vision Techniques.

Test No.	Measured ρ	Measured θ	Sonar ρ	Sonar θ	Vision ρ	Vision θ
1	919	-21	915.6	-20.6	888	-29.66
2	706	-27	715.4	-22.7	667	-35.51
3	930	20	924.0	23.2	783	23.99
4	1,242	0	1,226.3	4.6	1,128	10.27
5	764	32	778.5	46.1	593	43.62
6	1,164	-11	1,164.9	-13.7	1,084	-13.33
7	1,283	6	1,277.4	3.7	979	-6.53
8	1,319	-10	1,300.8	-9.8	1,084	-13.33

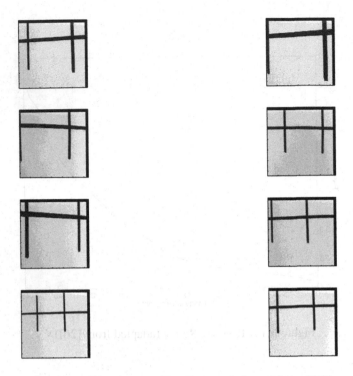

Figure 6.16: Visual Target Test Images (adapted from [26]).

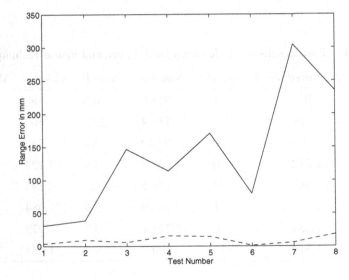

Figure 6.17: Error in ρ for Sonar (dashed line) and Vision (adapted from [26]).

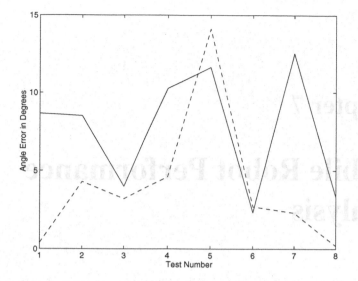

Figure 6.18: Error in θ for Sonar (dashed line) and Vision (adapted from [26]).

(For another detailed example comparing two alternative sonar sensor techniques to obtain wall pose, see [30].)

Note that, to keep things simple, we did not consider the error in the sonar location and orientation. However, these errors can be incorporated into the model in the same manner.

6.5   Conclusions

In this chapter, we have presented a theoretical framework for sensor modeling and design based on defining the performance semantics of the system. We introduced the notion of *instrumented sensor systems*, which is a modeling and design methodology that facilitates interactive, on-line monitoring for different components of the sensor system. It also provides debugging tools and analysis measures for the sensor system. The instrumented sensor approach can be viewed as an abstract sensing machine which defines the semantics of sensor systems. This provides a strong computational and operational engine that can be used to define and propagate several quantitative measures to evaluate and compare design alternatives. The implementation of this framework within the CWave system was described and examples were presented. This methodology is particularly appropriate for Computational Sensor Networks.

Acknowledgment

We would like to thank Professor Robert Kessler and Christian Mueller for providing the CWave program that we used to implement the instrumented sensor library, Professor Gary Lindstrom for his helpful discussions of program semantics, and Kevin Linen of North Carolina A & T for help with the experiments.

Chapter 7

Mobile Robot Performance Analysis

As stated in the introduction to the book, the CSN approach is based on the analysis of models of the sensor network, the physical phenomena, and the application scenario[1]. We apply this here to show that the exploitation of nonmobile, distributed sensor and communication devices by a team of mobile robots offers performance advantages in terms of speed, energy, robustness and communication requirements. At one extreme, mobile robots can be provided with a wealth of on-board sensing, communication and computational resources [8, 146]; at the other extreme, robots with fewer on-board resources can perform their tasks in the context of a large number of stationary devices distributed throughout the task environment [62]. In this study, all the models are simulated using software (C and Matlab), and the performance of robot tasks with and without the presence of an *S-Net* (i.e., a set of distributed sensor devices) is evaluated in terms of various measures.

The notions described above can be exploited in many situations and across several scales of application. Let us consider the following three: (1) fire fighting robots, (2) reservoir monitoring agents, and (3) wearable devices.

- Fire fighting:

    Suppose mobile robots are used to fight forest fires; then, there may be several hot spots to extinguish or attempt to control. If sensor devices can be distributed in the environment, then their values and gradients can be used to direct the behavior of fire fighting robots. Such mobile robots used as fire fighters can have several behaviors. They can transport fire extinguishing materials from a depot to the closest fire source and attempt to put out the fire. During this movement to and from the fire, collision avoidance algorithms can be employed. Sometimes coordinated activities are necessary and communication models are also important. Such a fire fighting behavior will continue until the current fire

---

[1]This chapter is a modified version of work done with Yu Chen [20].

T.C. Henderson, *Computational Sensor Networks*, DOI: 10.1007/978-0-387-09643-8_7,
© Springer Science+Business Media, LLC 2009

source is under control. Then the robots will move to the next serious source according to sensed temperature gradients.

- Reservoir task:

When swimming robots are used in a reservoir, some chemical sources can be detected and handled. If chemical concentration sensor devices can be distributed in the environment, their values and gradients can be used to direct the behavior of swimming robots. Such mobile robots can have several behaviors. They can transport neutralizer to the closest source, or they can block up the leaking source. During this process, coordinated activities and communication are necessary. The process will continue until the concentration in the whole reservoir is within a specified limit.

- Wearable devices:

Networked devices embedded in clothing or the external surface of a vehicle may be used to sense the environment and to automatically change the coloration of the clothing or vehicle to better suit a given task. For example, this could be used to blend into the background (i.e., camouflage), or to stand out from the environment (i.e., for rescue). The *S-Net* we are studying allow such capabilities.

7.1   Study Design

In this chapter, we provide models for various components of study: (1) mobile robots with on-board sensors (2) communication, (3) the *S-Net* (includes computation, sensing and communication), and (4) the simulation environment. We give algorithms developed for the *S-Net* which perform coöperative computations and provide global information about the environment using local and global frames as defined previously. The method for the production of global patterns using reaction-diffusion equations is exploited and its use for multi-robot coöperation demonstrated.

We describe the results of a set of experiments designed to help us better understand the benefits and drawbacks of *S-Nets*. For behaviors of one mobile robot going to a temperature source, and multiple mobile robots surrounding a temperature source, in the ideal situation, which means no noise is present, the *S-Net* takes more time and distance. But when noise is added in, which is more realistic, the *S-Net* is more robust than the *non-S-Net* system. For the last behavior of multiple mobile robots going back and forth to a temperature source, there are thresholds above which the *S-Net* system outperforms the *non-S-Net* system.

Several models are required in order to explore the questions that have been posed. These include a mobile robot model, sensor models, an *S-Net* model, a communication model, and a model of the environment. The simulation provides a computational framework for the interaction of these models in terms of mobile robots performing useful tasks in the environment, and we define our simulation model as well.

7.2 Mobile Robot Model

In order to act, a robot must receive current environmental information and calculate its movement based on the information received. On-board sensors (e.g., temperature, range, etc.) provide information about the environment and inform the robot's behaviors. In addition, the mobile robot may be able to communicate with other robots or the *S-Net*. The robot achieves movement by rotating or translating based on turning and motion primitives with given rotational and linear speeds.

A *Mobile Robot Model* is defined by:

- Local Frame:

    2D or 3D frame attached to the robot that provides the relative location of objects with respect to the mobile robot. An example of robot local frame is displayed in Figure 7.1.

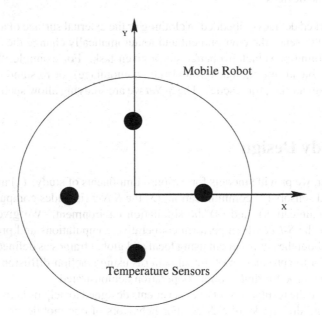

Figure 7.1: Local Robot Frame (adapted from [20]).

- Position Estimate:

    2D or 3D location estimate in world frame coördinates. It is used to control the robot's behavior.

- Heading Estimate:

    Orientation estimate in world frame coördinates. It is used to control the robot's behavior.

These are the robot's estimates, and may be different from the actual values in the environment model; this is caused by various sensor, actuation and computation errors.

- Description of On-board Sensors:

A simple distribution of on-board sensors is displayed in Figure 7.1. Four on-board temperature sensors are located on the axes of the local frame, and at a certain distance from the center of mobile robot. In the current implementation, the on-board sensors can be temperature sensors or range sensors.

- Primitive Behaviors:

Primitive motion functions available to the robot (e.g., turn, go forward, go backward, stop turn, stop go).

  { Turn: the robot can set its rotational speed as the maximum rotating speed in either the clockwise or counterclockwise directions.

  { Go Forward and Go Backward: the mobile robot can set the linear motion speed as maximum linear speed in either the positive or negative X direction of its local frame.

  { Stop: there are commands to stop rotation and stop linear motion.

- High-level Behaviors:

The high-level behavior of the robot is specified by a program which maps the robot state and environmental information to primitive behavior sequences. For example, the behavior for the mobile robot to go to the closest temperature source will include: mobile robot sensing to get environmental temperature and gradients, turning as well as going forward to the source, and finally stopping when it reaches a certain distance from the temperature source.

The *Environment Model* consists of:

- World Frame:

Base frame for world objects.

- Actual Robot Location:

This is the actual 2D or 3D position of robot in the world frame.

- Obstacles:

Location and shape of obstacles in the world frame. The central location of obstacles may be generated randomly by using a random number generator. The basic shape of obstacles will be square blocks with certain length edge. If two or more obstacles overlap, various shapes will form.

- Sources:

  Functions describing sources and distribution of energy, material, etc. (e.g., heat, chemicals, etc.). Multiple sources may be defined, the number and position of sources can be decided by the user. The formula for distribution of temperature is:

  $$T(x, y) = \frac{C}{\sqrt{(x - x_s)^2 + (y - y_s)^2 + 1}} \qquad (7.1)$$

  The temperature in a given location will be the maximum of all the temperature sources.

The *Sensor Model* is given by a specific model for each modality; this includes noise effects and others for each type of sensor. Here we specify models for the sensors used in our simulation. [Note: the robot can only estimate the actual environment variables' values by using its sensors.]

- Temperature Sensor Model:

  Ideally the sensors do not have any noise, but in practice, sensor data is corrupted by noise, e.g., Gaussian noise. A simple model for a temperature sensor is:

  $$\hat{T}(x, y) = T(x, y) + \mathcal{N}(\mu, \sigma^2) \qquad (7.2)$$

  where $T(x, y)$ is the actual temperature at location (x, y) in the environment and $\mathcal{N}(\mu, \sigma^2)$ is a normal distribution function with mean μ and variance σ^2. The sensor response may also be affected by nonlinear effects such as hysteresis or failure modes.

- Range Sensor Model (generic):

  A range sensor (e.g., sonar) can detect objects within a certain distance. Range sensor models depend on the sensor device geometry and physics as well as the structure of sensed surfaces, here we only give the generic form of expression:

  $$\hat{R}(x, y, \theta) = R(x, y, \theta) + \mathcal{N}(\mu, \sigma^2) \qquad (7.3)$$

  where $R(x, y, \theta)$ is the actual range of the nearest object from location (x, y) in direction θ, and $\mathcal{N}(\mu, \sigma^2)$ is Gaussian noise.

7.3   Communication Model

The *communication model* consists of a protocol, message layout, error model and performance characteristics. The protocol specifies the meaning of the bits in a message, as well as a set of commands for communication between robots and *SEL*s. A group of *SEL*s sharing a common frame is called an S-clique.

- Protocol:

Commands for a robot to communicate with *SELs*:

value	command	meaning
X0000000	reset	reset all the *SELs*
X0000001	is anybody there	robot requesting all *SELs* in range to respond
X0000010	talk to origin	robot communicates with origin of local frame to get information of *SELs* in local frame
X0000011	talk to one	robot communicates with *SELs* separately
X0000100	local position	robot communicates with origin of local frame, provides its position in local frame and asks origin for gradient

Commands for *SELs* to communicate with each other (suppose SEL_1 sends to SEL_2):

value	meaning
X1000000	reset all *SELs*
X1000001	SEL_1 requesting all *SELs* in range to respond (so distance can be calculated)
X1000010	SEL_1 responding to SEL_2 range request
X1000011	SEL_1 asserting it is origin in S-clique
X1000100	origin provides the robot gradient in local frame

In the simulation, the message sent out by robots or each *SEL* are renewed every hundredth of a second. The highest order bit (shown by the "X" in the command value is a bit used to specify whether it's a new message or not, i.e., if a robot or a *SEL* sends out a new message, the "X" bit in this agent's message is set to 1, otherwise it is 0.

Robots and *SELs* use the message structures described above in order to achieve various goals. For example, to determine a local frame, the following sequence is used:

{ Form local frame and find origin:
  * For each *SEL*, send command X1000001 to form an S-clique
  * After *SEL* receives other *SELs'* command to form S-clique, it responds to the ones within range
  * After finding the origins of local frames, they send out command X1000011 to assert that they are origin of S-clique
  * Next, the origin determines frame

- Message layout indicates the structure of each message sent out by either a mobile robot or a *SEL*. In the first byte of the message, there is a bit to indicate the new message, and a bit to indicate whether its a *SEL* or a robot. The other bits in this byte are reserved for future use. The next two bytes are the IDs for each agent. The bytes after depend on each command.

bytes\bits	7	6	5	4	3	2	1	0
1	new message	*SEL* or robot						
2			ID					
3								
			command dependent					

The command dependent part for robot communicates with the *SELs*:

{ Is Anybody There:

field No.	description	No. of bytes
4	empty	

{ Talk To Origin:

field No.	description	No. of bytes
4	SEL_1 ID	2
5	X position of SEL_1 in local frame	2
6	Y position of SEL_1 in local frame	2
7	SEL_2 ID	2
8	X position of SEL_2 in local frame	2
9	Y position of SEL_2 in local frame	2

{ Talk To One:

field No.	description	No. of bytes
4	number of agents want to talk with	1
5	agent 1 ID	2
6	agent 2 ID	2
.		
.		
.		

{ Local Position:

field No.	description	No. of bytes
4	*SEL* ID of origin	2
5	X position of robot in local frame	2
6	Y position of robot in local frame	2

{ Is *Home* Around:

field No.	description	No. of bytes
4	empty	

{ Talk to Blake Origin:

field No.	description	No. of bytes
4	ID of *SEL* want to talk to	

The command dependent part for robot communicates with robots:

field No.	description	No. of bytes
4	ID of the first cooperating robot	2
5	ID of the second cooperating robot	2
6	the distance to the first robot	2
7	the distance to the second robot	2

The command dependent part for *SEL*s communicate with each other or with mobile robots:

{ Command 0000000:

field No.	description	No. of bytes
4	empty	

{ Command 0000001:

field No.	description	No. of bytes
4	empty	

{ Command 0000010:

field No.	description	No. of bytes
4	agent ID of which asked	2
5	distance to the agent asked	2

{ Command 0000011:

field No.	description	No. of bytes
4	empty	

{ Command 0000011:

field No.	description	No. of bytes
4	robot ID	2
5	gradient in robot's position (local frame)	4

- Error:

The messages will be exchanged as packages. Messages can be lost during the process of communication; the effect of this error is that some agents do not receive the message. Various methods may be used to recover from this error (e.g., wait one communication period and let the source send the message one more time). Another kind of error is a wrong message, e.g., bad bits may be sent. Parity checking can be used to detect and correct this. For this study we explore only such simple kinds of errors and recovery.

- Performance:

The performance characteristics which can be set include:

{ bandwidth: bits per second transmission

{ range: maximum distance from device that signals can be received.

S-Net devices consist of three essential components: computation, sensing and communication. The *computation element* is described by the speed of the processor, its storage capacity, power requirements and cost. Sensors used by the *S-Net* devices are modeled as described above, but also include bandwidth, latency, power requirements and cost. The *communication model* is like that given for mobile robots, but includes power requirement and cost as well.

7.4   Simulation Model

We use discrete event simulation with a fixed time step. In that we must model and simulate continuous events (e.g., during robot motion) as well as discrete events, we allow for an *every-time-step* event which can be put at the head of the event queue and must be handled every time step. Any number of these may be added to the event queue.

The event list is a table recording all events that will happen in each time step. At the beginning of each time step, we copy it to a temporary list, and new events will be generated and added to the event list during the movement of the robots. The table has three fields:

- *Agent ID*: all the mobile robots and distributed sensors are agents with unique IDs. Robot IDs are distinguishable from *SEL* IDs.

- *Event type code*: indicates the type of behavior, including:

  { RunRobots: executes the robots' behaviors; this transforms the high-level behavior into a sequence of primitive behaviors such as: turn, go forward, go backward.

  { RunS-Elements: executes the *SELs*' behaviors; this includes communication between sensors and robot, forming local frames, and running local computations.

  { Turn, Go Forward, Go Backward: primitive behaviors for mobile robots.

  { Stop Turn: will stop the rotation of the robot.

  { Stop Go: will stop the linear motion of robot.

  { Broadcast: produces communication events.

- *Time code*: time of event execution. To handle continuous movements of mobile robots, some basic behaviors such as rotating and going will happen all the time, so we set up a special code which means to execute every time step. Other behaviors such as StopGo and StopTurn, are discrete and occur at the scheduled time.

During each time step, all scheduled events are handled and new events will be generated and added to the event list according to the different robot behaviors. At the end of each time step, the resulting state is evaluated to determine its feasibility. If an impossible state has occurred (e.g., a robot penetrates an obstacle), then a special handler is called to resolve the problem. Once a possible state is achieved, the status of each robot such as position and local direction is updated. This procedure is repeated until the simulation terminates.

The mobile robot's moving and turning behaviors are based on information provided by the *S-Net*. As an example, consider the case of temperature sensors. At one extreme, a mobile robot may have four on-board temperature sensors located in different positions, thus providing four different spatial samples. The temperature gradient can be determined from these four values. Equations [7.4] and [7.5] give the temperature function and gradient function, respectively. The temperature gradient can be used to control the heading of the robot (see Figures 7.2 and 7.3).

Figure 7.3 is a simulation of a mobile robot moving in the temperature gradient direction to find the temperature source. The mobile robot starts at the origin and moves to the temperature source (located at (-3,2)), which has the distribution function (7.4). The turning speed is π radians/sec and linear speed is 2 meters/sec. At the other extreme, with the *S-Nets*, the robot will obtain the gradient information from the scattered *SELs*.

$$T(x,y) = \frac{C}{\sqrt{(x - x_s)^2 + (y - y_s)^2} + 1} \tag{7.4}$$

$$\nabla T = \begin{bmatrix} \frac{\partial T}{\partial x} \\ \frac{\partial T}{\partial y} \end{bmatrix} = \begin{bmatrix} \dfrac{-C(x-x_s)}{\sqrt{(x - x_s)^2 + (y - y_s)^2}(1 + \sqrt{(x - x_s)^2 + (y - y_s)^2})^2} \\ \dfrac{-C(y-y_s)}{\sqrt{(x - x_s)^2 + (y - y_s)^2}(1 + \sqrt{(x - x_s)^2 + (y - y_s)^2})^2} \end{bmatrix} \tag{7.5}$$

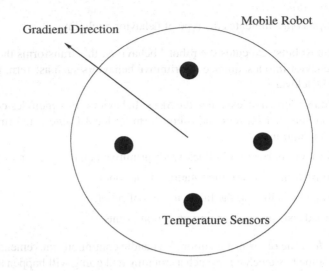

Figure 7.2: Gradient Calculation and Following (adapted from [20]).

When more than one robot is in use and without *S-Nets*, direct coöperation between robots is necessary. A set of algorithms are needed to prevent robots' collision and to improve the operational efficiency (e.g., minimum total distance, etc.). Our goal is to study the various aspects of the relationship between robots with and without *S-Nets*. Variables of interest include: sensor distribution, robot and *SEL* parameters, and sensor performance.

When using multiple robots and devices, communication between agents plays an important role. The robots should be able to communicate (e.g., current position, speed, etc.) with each other, as well as *SELs* and external controllers. A simple and efficient communication protocol has been described in Section 2.1.

We study robot performance in terms of behaviors developed for various scenarios including both military (e.g., tactical urban settings) and disaster mitigation (e.g., chemical spills, forest fires). Robots can be used to control fire or detect poisonous gas sources; i.e., things that are difficult and dangerous for human beings.

7.5 Goal Achievement

The most important criterion for robot evaluation is the successful achievement of its goals. Assuming that various strategies are all successful, the next level of comparison is achieved in terms of the efficiency of the behavior. A basic set of goals for mobile robots are:

- *Go to geometric destination:* This may involve either absolute or relative locations as well as the ability to maintain relative positives (e.g., follow another robot).

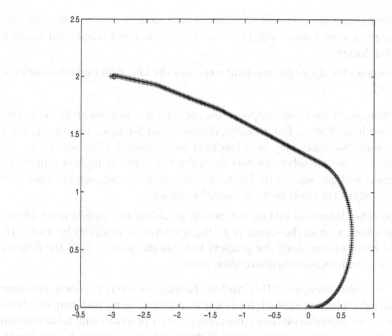

Figure 7.3: Simulation of a Mobile Robot Moving against Temperature Gradient (adapted from [20]).

{ *Go to absolute location:* Suppose we know the global position of the mobile robot and the *SEL*s and all the agents belong to the same global frame, then *go to absolute location* can be implemented directly.

{ *Go to relative location:* In most cases, a global frame may not be available. Information of the environment can be obtained only through local frames. The location of destination is relative to the local frame, and the mobile robot needs to exploit the transformation between local frames to reach the final destination.

- *Go to destination related to source:* This involves moving with respect to the sources of interest in the particular problem (e.g., a temperature source). Moreover, this may be based on the known values at the *SEL*s or interpolated values.

  For those behaviors that utilize the *S-Net*, all the information about source comes from *SEL*s, therefore the mobile robot needs to communicate with the origins of various S-cliques or even with particular *SEL*s (e.g., the *SEL* with highest temperature in some S-clique). To achieve this, the robot:

  { Communicates with the origin and the other two *SEL*s in an S-clique frame to decide its current position in that frame.

  { After moving a certain distance, the robot again communicates with them to determine the new location.

{ Transforms the pertinent information such as the location of the *SEL* with highest temperature value to its own frame (the robot frame) and moves to that location.

{ Repeats the above process until it reaches the *SEL* with highest value of all *SELs*.

In the discussion above, we suppose there are at least two common *SELs* between every two local frames. But in reality, disconnected S-cliques may occur, which means some S-cliques may have less than two common *SELs* with others. In this case, the mobile robot can only reach the *SEL* with the highest value in the connected S-clique set. To do better, it is necessary to increase the number of *SELs*, or adjust the range of the S-clique broadcast.

For the robot behaviors that do not exploit an *S-Net*, the mobile robot obtains the information about the source (e.g., the temperature gradient) by itself. The mobile robot moves along the gradient towards the source, until the detected value (e.g., temperature) is above some limit.

- *Go along constrained path:* This involves the incorporation of various constraints into the path selection method; for example, the least costly path may be desired (cost may be related to distance, time, energy, etc.) or a path with some constant value in a space of interest (e.g., constant distance from a source), or may involve other desired properties (e.g., avoid collisions).

7.6   Multiple Robot Behaviors

Here we consider two basic behaviors:

- Multiple mobile robots cooperate and communicate via the *S-Net* to find temperature source and keep certain distance from the temperature source evenly.

When the robots reach a certain distance away from the highest temperature *SEL*, robots communicate with each other to get their distances. Then two of the mobile robots that are close to each other keep their positions and the farthest one from them computes the positions that form an equilateral triangle; from the two results, the robot chooses the closer one and moves to it.

- Multiple mobile robots go back and forth to a temperature source, the intensity of the temperature source will decrease after each robot's visit, and finally the temperature of whole environment area will be controlled.

Home is chosen as the origin of the S-clique that sensed the lowest temperature, then stripe patterns are formed along the gradient of the temperature source to *Home*. The straight line from *Home* to the temperature source is in the middle of a white stripe, and black stripes alternate spatially with white stripes. The width of each stripe is a constant. The robots will move from *Home* to temperature source along white stripe and follow the black stripe back *Home*. During the procedure, if any robot detects that a collision is about to happen, it will slow down to avoid the collision.

The development up to this point has created the framework in which the performance of mobile robots can be compared with respect to using the *S-Net* or not. We now compare the performance of mobile robots while solving the following tasks:

- One robot goes to a temperature source.

- Multiple robots surround a temperature source.

- Multiple robots go back and forth to a temperature source.

This set of tasks represents typical mobile robot tasks and can be configured to exploit many of the constraints described in earlier chapters. For example, a robot's path may be required to be the shortest, the gradient may be followed, or patterns in the *S-Net* may be used as road markers. Moreover, the last two tasks provide a setting to use multiple robots, ranging from few to many robots. In addition, robot interactions are necessary, at least as far as avoiding collisions. For each of these tasks, we propose a relevant set of performance measures, as well as a discussion of parameters and their possible values. Finally, we give the performance results and compare the two approaches.

Our goal is to find out under what conditions, the *S-Net* system can perform better or more robustly and the cost by analyzing the measurement such as time used, distance traveled and the final distance to the temperature source. According to the final result, for the first two behaviors, the *S-Net* system does not perform better when no noise is present, but when the noise is added in, we find that the *S-Net* system is more robust, especially in rough situations, in which there is lots of noise in the environment. For the third behavior, the *S-net* system not only performs much better under realistic situations, but even under ideal conditions, it displays a benefit. For a certain distance of round trip, the *s-Net* system can support more robots and prevent collisions happen between each other. On the other hand, if there are too many robots in the *non-S-Net*, some robots cannot move properly and prevent collision.

Overview of methodology:

- One robot goes to a temperature source

  { The system without the *S-Net*:
  The robot uses four on-board temperature sensors to detect the temperature and moves along the temperature gradient toward the temperature source.

  { The *S-Net* system:
  The robot communicates with the *SEL*s around and finds out the closest origin of an S-clique; then the robot communicates with this origin to transform the position of the *SEL* with highest temperature in the S-clique to its own frame, and move to that position. This procedure repeats until the robot reaches the highest *SEL* of the entire *S-Net*.

- Multiple robots surround a temperature source

  { The system without the *S-Net*:
  Three robots begin from different places, and each of them uses four on-board temperature sensors to detect the temperature and move along the

temperature gradient toward the temperature source. When the average temperature detected is above a certain value, they stop and try to communicate with each other. Then one robot will maintain its position, and the other robots will move following a constant-valued temperature contour. To do this, they compute the gradient and move perpendicular to it.

{ The *S-Net* system:
Three robots begin from different places, and each of them uses local frame transformations to move to the *SEL* with the highest temperature. When all robots reach a certain distance away from the highest temperature *SEL*, two of the robots keep their positions, and the other one which is the farthest from them will move forward to form an equilateral triangle.

- Multiple robots go back and forth to a temperature source

{ The system without the *S-Net*:
Each robot moves from *Home* using four on-board temperature sensors to detect the temperature and moves along the temperature gradient toward the temperature source. Then they turn around, using four on-board *Home* sensors to move back to *Home*. The robot behavior to prevent collision is to detect a collision, make a right turn and then try to get back on track again.

{ The *S-Net* system:
The stripe patterns are formed along the gradient of the temperature source to *Home*. Each robot begins from the same place and moves along the white stripe toward the temperature source and follows the black stripe back *Home*. When a robot detects that a collision is about to happen, it will slow down to prevent the collision, until it cannot detect the collision any more.

7.7   One Robot Goes to a Temperature Source

This is the simplest behavior for a robot, and the major focus of this experiment is to analyze the robustness of the system by changing the parameters of the temperature source and controlling the noise parameters in the sensors.

We compare the performance of:

- One robot without on-board sensors utilizing the *S-Net* to reach the closest temperature source.

- One robot using only on-board sensors to reach the closest temperature source.

Examples of these two alternatives are shown in Figures 7.4 and 7.5, in which there is one temperature source ("◇") in the environment. Figure 7.4 is an isotherm plot describing the distribution of temperature, in which a robot starts at the origin $(0, 0)$ and follows the temperature gradient to reach the source, which is located at $(3, 2)$. In order to make the isotherm plot more clear, we amplify the units by 10. Figure 7.5 not only gives the trace of a robot goes to the destination with the *S-Net*, but also provides the

formation of local frames. In the figure, each "+" is one *SEL*, and each "*" is the origin
of one local frame, and "– – –" is the range of local frame origin. As we mentioned
before, there should be at least two common *SELs* between any two close frames in
order to transform coördinates between them. In Figure 7.5, a robot also starts at the
origin (0, 0) and moves to the temperature source (3, 2) by using local S-clique frames.
Since by using the *S-Net*, the robot can only estimate the destination location in its own
frame, as well as the fact that the robot has a non-zero turning radius, this causes the
robot to go to a place close to the temperature source instead of exactly to the source.

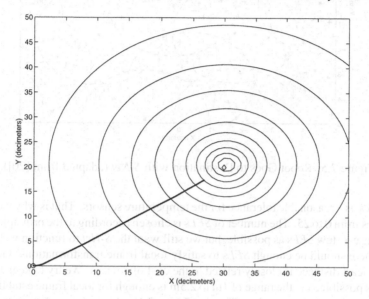

Figure 7.4: Robot Goes to Destination without *S-Net* (adapted from [20]).

Constants set up for the experiments are:

Maximum Linear Speed	1 m/s
Maximum Rotation Speed	π rad/s
Initial Location of Robot	(0, 0) ("o" in the figure)
Initial Direction of Robot	0 rad
Number of Sources	1
Location of Source	(3, 2)

in which the initial location and initial direction of robot, as well as the location of
sources are all according to home frame of environment instead of any local frame.

In these experiments we test performance time and distance traveled with respect
to sensor noise or variance (0 to 25), number of *SELs* (100 to 300), and broadcast
distance (1 to 2.5) for the *SELs*. According to [41, 152, 37], noise of a sensor includes
inherent noise, transmitted noise, mechanical noise and setback noise and so on. The
temperature sensor model we choose here has the range of $[0, 1000^oC]$, and we think

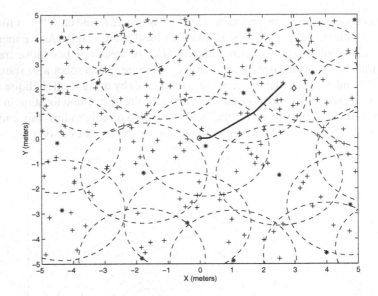

Figure 7.5: Robot Goes to Destination with *S-Net* (adapted from [20]).

that 0.05% is a reasonable tolerance for the temperature sensors. That is why we choose σ^2 ranges from 0 to 25. The number of *SEL*s is chosen according to the principle that we want to use as few *SEL*s as possible, but we still want the *S-Net* to function well, which means there should be enough *SEL*s to satisfy local frame transformations. Generally, the broadcast distance is closely related to the cost of devices. We try to keep the price as low as possible, and the range of 1m to 2.5m is enough for local frame establishment.

The results are displayed in Figures 7.6 to 7.18. Each data point represents the estimated value of the performance measure of interest, and is the mean of ten simulation experiments. The variance is also shown. The stochastic part of each experiment is the location of the *SEL*s. The reason we choose ten simulation experiments is according to the confidence interval. Suppose we want to obtain an approximate 90% confidence interval for the expected average time utilization, which is given by:

$$E(X) = E(\frac{\sum_{i=1}^{N} D_i}{N})$$

From the ten replications we obtain:

$$\overline{X}(10) = 4.95$$

$$S^2(10) = 1.34$$

and the confidence interval is:

$$\overline{X}(10) \pm t_{n-1,1-\frac{\alpha}{2}} \sqrt{\frac{S^2(n)}{n}}$$

$$= \overline{X}(10) \pm t_{9,0.95} \sqrt{\frac{S^2(10)}{10}} = 4.95 \pm 0.679$$

Thus, subject to the correct interpretation to be given to a confidence interval, we can claim with approximately 90% confidence that $E(X)$ is contained in the interval $[4.271, 5.629]$ seconds.

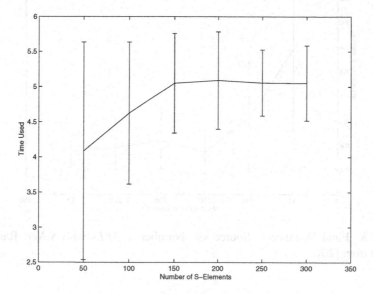

Figure 7.6: Time vs. Number of *SEL*s with *S-Net*; Range = 2 (adapted from [20]).

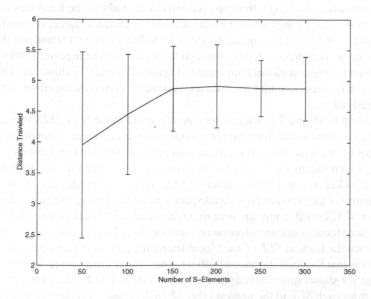

Figure 7.7: Distance Traveled vs. Number of *SEL*s with *S-Net*; Range = 2 (adapted from [20]).

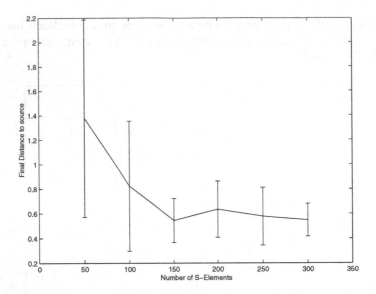

Figure 7.8: Final Distance to Source vs. Number of *SELs* with *S-Net*; Range = 2 (adapted from [20]).

The robot first communicates with the S-clique origin with the highest temperature within communication range, then gets the information about the local frame from it. After the robot determines its own position in the S-clique's frame, it transform the location of the *SEL*, with the highest temperature value, to its own frame and then goes to that location. This process repeats through S-cliques within range until robot reaches the *SEL* with the highest global temperature. Figures 7.6 and 7.7 show that more *SELs* will cause an increase in both time used and distance traveled by the robot. This effect is explained below.

According to Figure 7.8, we can see that when the number of *SELs* increases, the final distance of the robot from the temperature source decreases dramatically. When the number of *SELs* increases, the variances of both time utilized and distance traveled decrease, which means the system becomes more robust. The fact that time used and distance traveled increase with number of *SELs* relates to the density of the *SEL* set and the ability of the *S-Net* to provide adequate spatial resolution. Because the increase of number of *SELs* will cause an increase in the number of S-cliques and local frames formed, more local frame transformations will be done by the robot. When the robot tries to reach the highest *SEL* of each local frame the path becomes more zigzag. But the payoff is that the final distance of robot to the temperature source decreases.

Figure 7.9 shows quantitatively that as the number of *SELs* increases, the average distance from each *SEL* to the nearest other *SELs* decreases. So suppose the robot can reach the highest *SEL* of the whole system, when given a specified accuracy for the final location with respect to a temperature source, we can determine the number of *SELs* required. For example, to guarantee that a mobile robot can reach a temperature

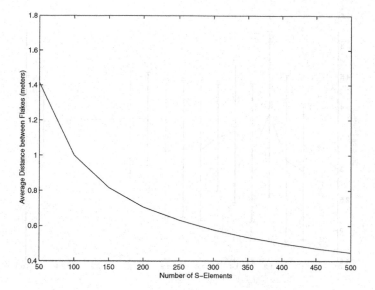

Figure 7.9: Average Distance of *SEL*s vs. Number of *SEL*s in the Area (adapted from [20]).

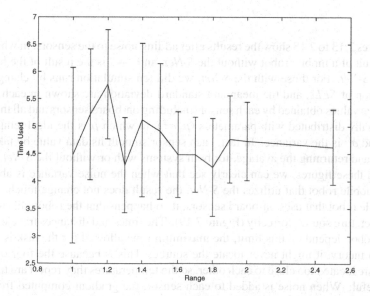

Figure 7.10: Time vs. Range with *S-Net*; Number of *SEL*s = 250 (adapted from [20]).

source within 1 meter, we need at least 100 *SEL*s over an area about 10*m* ∗ 10*m* (the radius of each robot is 0.1m).

According to Figures 7.10, 7.11, and 7.12, we can see that when the radio broadcast range is less than 1.5m, the behavior is less robust (variances are large). When the range is between 1.5 and 2.0m, the variance decreases and the behavior becomes more robust.

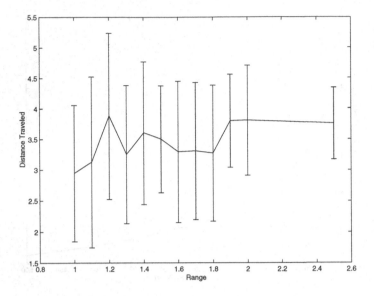

Figure 7.11: Distance Traveled vs. Range with *S-Net*; Number of *SELs* = 250 (adapted from [20]).

Figures 7.13 to 7.18 show the results after adding noise to the sensors, in which "..." is the result of a mobile robot without the *S-Net*, and "—" is the result of the behavior with the *S-Net*. For those with the *S-Net*, we did ten simulation runs by changing the distribution of *SELs*, and the mean and standard deviation are shown in each sample point. The values obtained by each sensor (including on-board sensors and all the *SELs*) are normally distributed with parameters (μ, σ^2), in which μ is the ideal temperature value, and σ^2 is the variance. In fact, each sensor smooth its data value by taking ten samples and returning the average, for both systems with or without the *S-Net*.

From these figures, we can clearly see that when the noise variance is above 10, for the mobile robot that utilizes the *S-Net*, the result does not change much. But for the mobile robot that uses on-board sensors, it so happens that the robot fails to locate the temperature source correctly (Figure 7.19). The times and distances traveled by the mobile robot depend on this limit; the maximum time allowed for the task is 15 time units. In theory, it might never locate the source. This is because the four on-board sensors are located too close to each other, so the temperatures they report are too noisy to be useful. When noise is added to each sensor, the gradient computed from their values can have large error, which will further change the direction the mobile robot moves. One proposed solution to this problem is to have the mobile robot move to four widely spaced locations and get samples across a greater spatial scale to compute the correct gradient. This will certainly cost much in time and energy. In fact, it also reduces the accuracy with which the robot can locate the source.

For some specific cases, e.g., there are two or more temperature sources in the environment, and mobile robot located in the exact middle area of the sources, it is difficult for the mobile robot to locate the source when noise is added to the sensors.

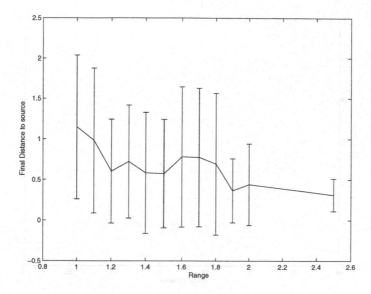

Figure 7.12: Final Distance to Source vs. Range with *S-Net*; Number of *SELs* = 250 (adapted from [20]).

Figures 7.21 and 7.22 show the results of this experiment, with noise variance $\sigma^2 = 8$, the mobile robot with on-board sensors cannot figure out the exact location of the temperature source.

From all these measurements and comparisons of the two systems, we can see that when in the ideal situation, which means no noise, the *S-Net* takes more time and distance. Compared to the *non-S-Net* (time used averages 3.22sec, distance traveled averages 3.21m), the cost of time ranges from 3.6(sec) to 5.5(sec), and distance traveled from 3.6m to 5.2m. But when noise is added in, which is more realistic, the *S-Net* system basically does not change much, but the system without the *S-Net* gets much worse. So we conclude that when in real situations, especially tough situations with lots of noise, the *S-net* system will be more robust than the system without the *S-Net*.

7.8   Multiple Robots Surround Temperature Source Evenly

This experiment is designed to explore the benefits of using the *S-Net* with regard to multiple mobile robot coöperation. Also, we use the same behavior in all the robots, so that by satisfying the same set of constraints, the robots can achieve the desired final result.

We compare the performance of:

- Three mobile robots without on board temperature sensors utilizing the *S-Net* to surround the *SEL* with the highest temperature value.

- Three mobile robots using only on-board temperature sensors to surround the temperature source evenly.

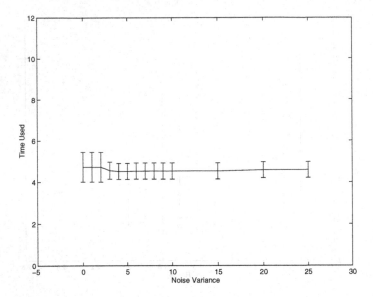

Figure 7.13: Time vs. Variance of Noise With *S-Net*; Number of *SEL*s = 300, Range = 2 (adapted from [20]).

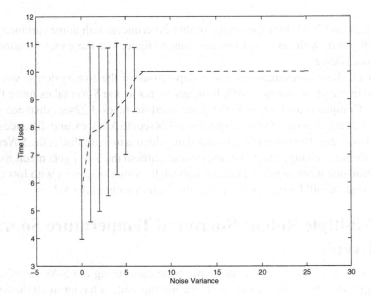

Figure 7.14: Time vs. Variance of Noise Without *S-Net* (adapted from [20]).

Examples of these two alternatives are shown in Figures 7.23 and 7.24, in which three mobile robots originated from different places ("o"), and there is a temperature source ("⋄") in the environment. "*" in Figure 7.23 is the position of the *SEL* with highest temperature.

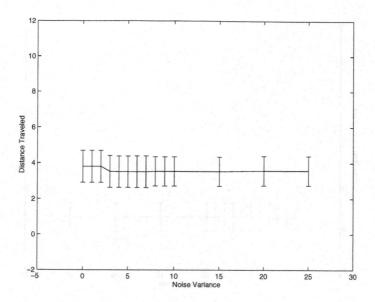

Figure 7.15: Distance Traveled vs. Variance of Noise With *S-Net*; Number of *SEL*s = 300, Range = 2 (adapted from [20]).

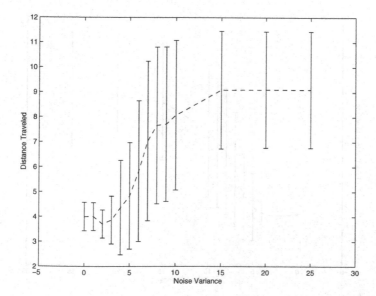

Figure 7.16: Distance Traveled vs. Variance of Noise Without *S-Net* (adapted from [20]).

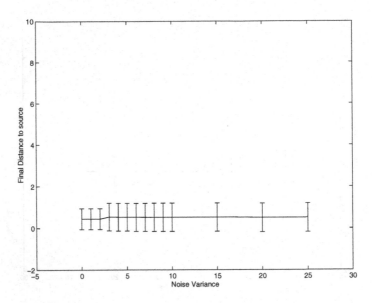

Figure 7.17: Final Distance to Source vs. Variance of Noise With *S-Net*; Number of *SEL*s = 300, Range = 2 (adapted from [20]).

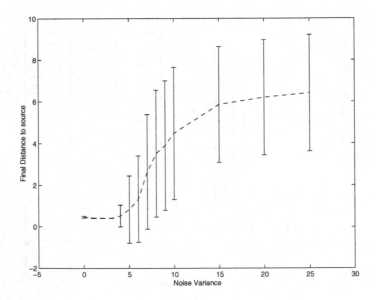

Figure 7.18: Final Distance to Source vs. Variance of Noise Without *S-Net* (adapted from [20]).

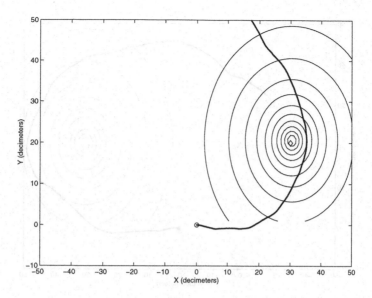

Figure 7.19: Robot Goes to Destination without *S-Net*; with noise variance = 10 (adapted from [20]).

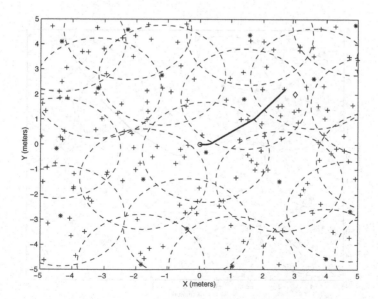

Figure 7.20: Robot Goes to Destination with *S-Net*; with noise variance = 10 (adapted from [20]).

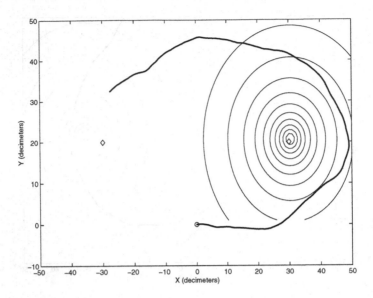

Figure 7.21: Robot Goes to Destination without *S-Net*; with noise variance = 8 (adapted from [20]).

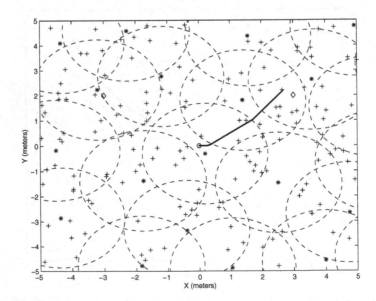

Figure 7.22: Robot Goes to Destination with *S-Net*; with noise variance = 8 (adapted from [20]).

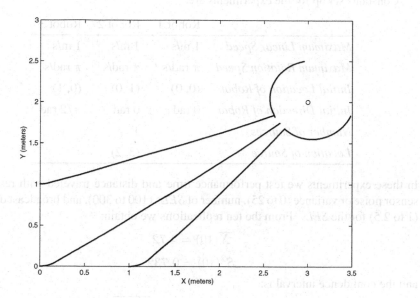

Figure 7.23: Three Robots Surround the Temperature Source Without *S-Net* (adapted from [20]).

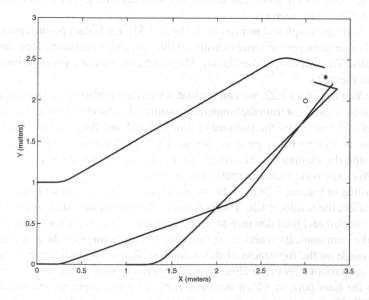

Figure 7.24: Three Robots Surround the Highest *SEL* with *S-Net* (adapted from [20]).

Constants set up for the experiments are:

	Robot 1	Robot 2	Robot 3
Maximum Linear Speed	1 m/s	1 m/s	1 m/s
Maximum Rotation Speed	π rad/s	π rad/s	π rad/s
Initial Location of Robots	(0, 0)	(1, 0)	(0, 1)
Initial Direction of Robot	0 rad	0 rad	$\pi/2$ rad
Number of Sources		1	
Location of Source		(3, 2)	

In these experiments we test performance time and distance traveled with respect to sensor noise or variance (0 to 25), number of *SEL*s (100 to 300), and broadcast distance (1 to 2.5) for the *SEL*s. From the ten replications we obtain:

$$\overline{X}(10) = 4.72$$

$$S^2(10) = 0.732$$

and the confidence interval is:

$$\overline{X}(10) \pm t_{n-1,1-\frac{\alpha}{2}} \sqrt{\frac{S^2(n)}{n}}$$

$$= \overline{X}(10) \pm t_{9,0.95} \sqrt{\frac{S^2(10)}{10}} = 4.72 \pm 0.496$$

Thus, we can claim with approximately 90% confidence that $E(X)$ is contained in the interval $[4.224, 5.216]$ seconds.

The results are displayed in Figures 7.25 to 7.31. Each data point represents the estimated value of the performance measure of interest, and is the mean of ten simulation experiments. The variance is also shown. The stochastic part of each experiment is the location of the *SEL*s.

According to Figure 7.27, we can see that when the number of *SEL*s increases, the final distance of the robot from the temperature source decreases dramatically. When the number of *SEL*s increases, the variance of time utilized and distance traveled decrease; this means the system becomes more robust. The fact that time and distance traveled increase with the number of *SEL*s relates to the density of the *SEL* set and the ability of the *S-Net* to provide adequate spatial resolution.

According to Figures 7.28 to 7.29, we notice that in this particular behavior, range does not affect the results much. This is because the parameters, such as time utilized, distance traveled and final distance to source, depend not only on each specific robot, but also the communication and coöperation of the three robots, which decreases the effect of range on the robustness of this behavior. Range can only control the aspect in which each robot tries to get close to the temperature source, but does not have any effect on the later part, in which the three robots try to cooperate and surround the highest *SEL*. However, both parts account for the time used and distance traveled. It is sure that the final distance to the source is decreased, and this is significant because of the effect of the range parameter.

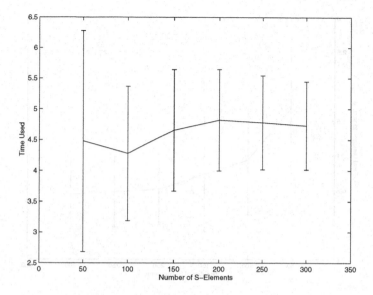

Figure 7.25: Time vs. Number of *SEL*s for Robots with *S-Net*; Range = 2 (adapted from [20]).

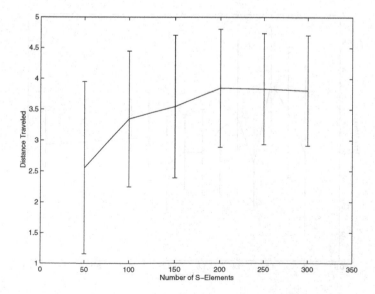

Figure 7.26: Distance Traveled vs. Number of *SEL*s for Robots with *S-Net*; Range = 2 (adapted from [20]).

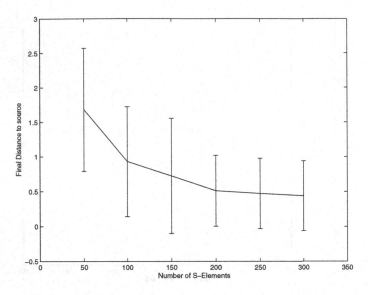

Figure 7.27: Final Distance to Source vs. Number of *SEL*s for Robots with *S-Net*; Range = 2 (adapted from [20]).

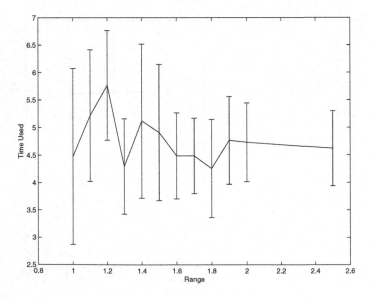

Figure 7.28: Time vs. Range for Robots with *S-Net*; Number of *SEL*s = 300) (adapted from [20]).

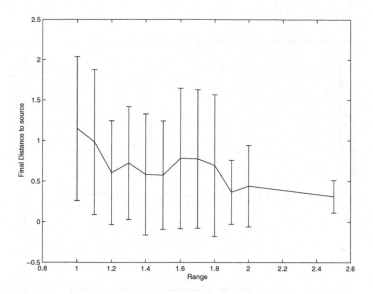

Figure 7.29: Final Distance to Source vs. Range for Robots with *S-Net*; Number of *SELs* = 300 (adapted from [20]).

Figures 7.30 to 7.31 show the results of adding noise to sensors, in which "..." is the result of mobile robots without the *S-Net*, and "—" is the result of the behavior with the *S-Net*.   For those with the *S-Net*, we did ten simulation runs by changing the distribution of *SELs*, so the mean and standard deviation are shown in each sample point. The values obtained by each sensor (including on-board sensors and all the *SELs*) are normally distributed with parameters (μ, σ^2), in which μ is the ideal temperature value, and σ^2 is the variance. In fact, each sensor smooth the data value by taking ten samples and returning the average.

From these figures, we can clearly see that, when the noise variance is above 10, for the mobile robot that utilizes the *S-Net*, the result does not change much. But for the mobile robot that uses on-board sensors, the robot can not surround the temperature source correctly (Figure 7.32).

In the robot system without the *S-Net*, when robots reach a certain distance away from the temperature source, which means when their sensed temperatures are above some level, the robots will begin to cooperate. One robot, which is closer to the others will maintain its position, and the other robots will move following a constant-valued temperature contour. To do this, they compute the gradient and move perpendicular to it. While tracking the contour, even a small amount of noise will cause them to move away from the contour, and results in their inability to finish the surrounding task. On the other hand, in the *S-Net* system, when robots reach a certain distance away from the highest temperature *SEL*, two of the robots keep their positions, and the other one which is farthest from them will move forward to form an equilateral triangle. So we conclude that when in real situations, especially tough situations with lots of noise, the *S-Net* system will be more robust than the system without the *S-Net*.

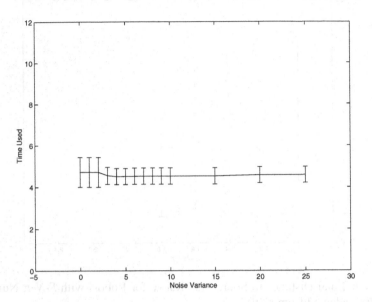

Figure 7.30: Time vs. Variance of Noise for Robots with *S-Net*; Number of *SELs* = 300, Range = 2 (adapted from [20]).

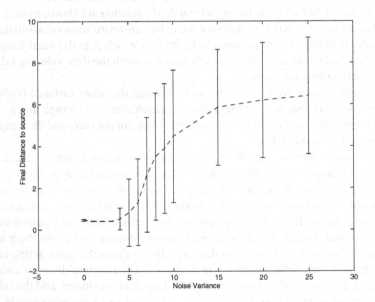

Figure 7.31: Final Distance to Source vs. Variance of Noise for Robots Without *S-Net* (adapted from [20]).

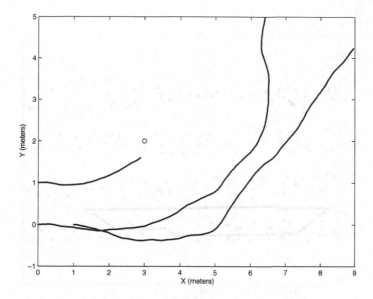

Figure 7.32: Robots Surround the Temperature Source without *S-Net*; with noise variance = 10 (adapted from [20]).

7.9 Multiple Robots Go Back and Forth to the Temperature Source

This experiment is also designed to explore the benefits of using the *S-Net* with regard to multiple coöperating mobile robots. Also, we would like to use the same behavior in all the robots, so that by satisfying the same set of constraints, the robots can achieve the desired final result.

We compare the performance of:

- Multiple mobile robots without on-board temperature sensors which utilize a stripe pattern formed by the *S-Net* to go back and forth to the temperature source from a home location.

- Multiple mobile robots using only on-board temperature sensors and *Home* sensors to go back and forth to the temperature source from a home location.

Examples of these two alternatives are shown in Figures 7.33 and 7.34, in which mobile robots originated from *Home* ("o" in figures), and there is a temperature source ("⋄" in figures) in the environment.

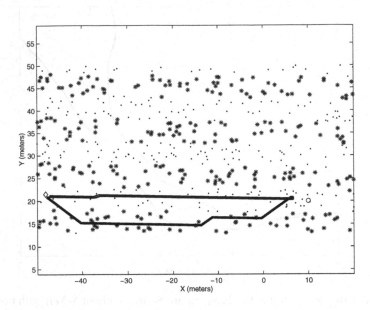

Figure 7.33: Trace of Robots Going Back and Forth with *S-Net*; 2 robots, 2 round trips (adapted from [20]).

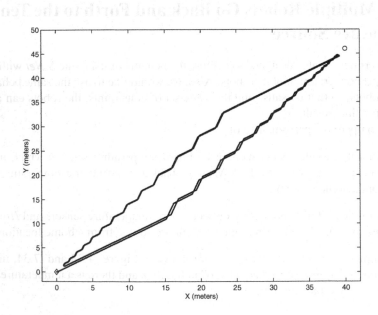

Figure 7.34: Trace of Robots Going Back and Forth without *S-Net*; 2 robots, 2 round trips (adapted from [20]).

Constants set up for the experiment are:

	Robot
Maximum Linear Speed	10 m/s
Maximum Rotation Speed	20π rad/s
Initial Direction of Robot	0 rad
Number of Sources	1

In the case that mobile robots use the *S-Net* (500 *SELs*), *Home* is chosen as the origin of the S-clique that sensed the lowest temperature; this provides the longest path to the maximum temperature *SEL*. Then stripe patterns are formed along the gradient of the temperature source to *Home*. The straight line of *Home* to temperature source (located in [10, 20]) is in the middle of the white stripe (pattern value is 1), the width of each stripe is a constant (5m in our case), and black stripes (pattern value is 0) alternate spatially with white stripes. Different stripe patterns are formed for different random streams. The robots will move along the white stripe toward the temperature maximum and follow the black stripe *Home*. When a robot detects that a collision is about to happen, it will slow down to prevent the collision.

In the case that mobile robots do not use the *S-Net*, *Home* is arbitrarily located at the origin of environment (0, 0), and the temperature source is located according to the average distance of *Home* to the temperature source in the *S-Net* experiments. The purpose of this is to make sure that the distances of the round trip are basically the same for both setups. We believe that the gradient of temperature source to *Home* does not affect our experiment, so we let the temperature source be located in (40, 46). When robots detect an environment collision, they make a right turn and then try get back on track again. The robots do not use the method of slowing down, because it is difficult and time consuming to determine whether the robot in front is moving toward them or moving in the same direction as they are moving.

In these experiments we test performance time and distance traveled, with respect to the number of round trips (1 to 10) and the number of robots on the path (1 to 10). The results are displayed in Figures 7.35 to 7.38.   Each data point represents the estimated value of the performance measure of interest, and is the mean of twelve simulation experiments. The stochastic part of each experiment is the location of the *SELs*.

Figures 7.35 and 7.36 give the performance of the *S-Net* system, from which we can see that when the number of robots and trips increase, the average time used and distance traveled by each robot increases linearly, and there are no major deviations from linear. When we take a close look at the data collected, we find that on occasion, due to the particular random number streams, the result is not ideal, which means the robots cannot exactly follow the stripe, but get lost looking for the correct stripe. Under detailed analysis, we found that it is caused by some particular distributions of *SELs*. Since we use the origin with lowest temperature as the *Home*, it is sometimes possible that it is on the border of the stripe. There may not be enough *SELs* on the border for black stripes, and then when the robots try to follow the stripe to go *Home*, they may not get enough information to keep on the black stripe, and thus move away. This can

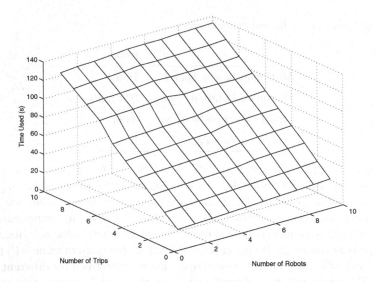

Figure 7.35: Time Used by up to 10 Robots for up to 10 Round Trips with *S-net*; number of *SEL*s = 500 (adapted from [20]).

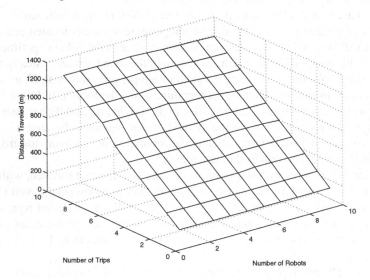

Figure 7.36: Distance Traveled by up to 10 Robots for up to 10 Round Trips with *S-net*; number of *SEL*s = 500 (adapted from [20]).

be solved by making more *SEL*s on the border or making *Home* far away from borders. But this is just some particular cases, which can be handled accordingly in reality, and is not unsolvable.

Figures 7.37 and 7.38 give the performance of the system without the *S-Net*, from which we can see that when the number of robots and trips increased, the average time used and distance traveled by each robot increase linearly. After a detailed analysis of

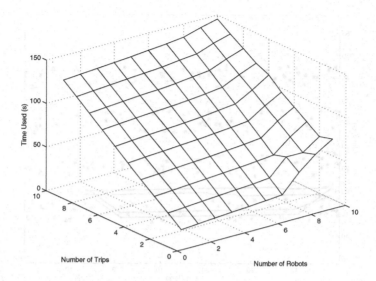

Figure 7.37: Time Used by up to 10 Robots for up to 10 Round Trips without *S-Net* (adapted from [20]).

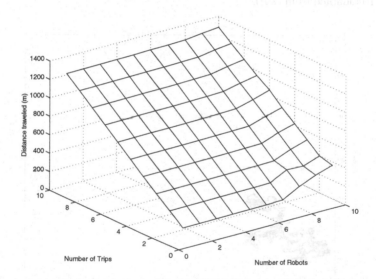

Figure 7.38: Distance Used by up to 10 Robots for up to 10 Round Trips without *S-Net* (adapted from [20]).

the data collected, we found that when there are more than eight robots on the same path, several robots may lose control. This is related to the robot behavior chosen. While there are lots of other behaviors, we believe that this is a rather standard collision avoidance algorithm and representative of many implementations in physical systems.

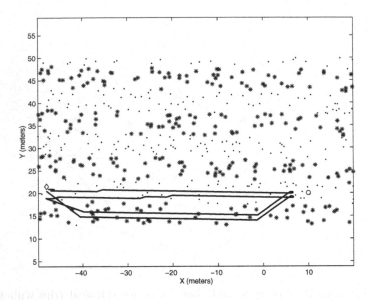

Figure 7.39: Trace of Robots Going Back and Forth with *S-Net*; 2 robots, 2 round trips, noise = 10 (adapted from [20]).

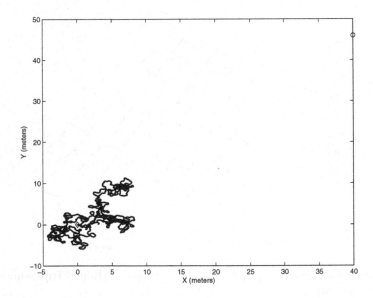

Figure 7.40: Trace of Robots Going Back and Forth without *S-Net*; 2 robots, 2 round trips, noise = 0.5 (adapted from [20]).

Figures 7.39 and 7.40 show how noise affects the performance in both cases. From these figures, we found that the one using *S-Net* can handle noise very well, when the noise variance is about 10, the performance and trace of robots generally stay the same. But for the case that does not use the *S-Net*, noise variance has a huge effect on the performance of the robots, where even a tiny variance as little as 0.5 can cause the robots to lose control. So we can conclude that, in terms of noise, the *S-Net* performs much better.

The performance measures used to this point have looked at success/failure, time to goal and distance traveled. Another crucial aspect is the more qualitative users' defined cost which is, in general, a function of the physical performance measures. For example, it may be that timeliness is extremely important, and the user may assign an exponential cost to time. Even if the cost is directly proportional, it may be linearly related with a steep slope.

To explore this aspect of performance cost, we have set up two models: (1) linear, and (2) quadratic. The three major terms included are:

- robot cost: we always assume this is linear in the number of robots.

- *S-net* cost: we always assume this is linear in the number of *SEL*s.

- physical quantitative (e.g., time and distance determined from simulation experiments): we apply a linear or quadratic forum to this term.

In order to explore this issue, we examined linear and quadratic cost functions in terms of parameters in the equations in order to determine the existence of various cost requires related to parameter values.

We define the cost relation as:

$$C_l = C_s + C_p$$

where:

- C_l is the total cost

- C_s is the cost of the system infrastructure

- C_p is the cost of performance

- $C_s = N_r * C_r + N_{s-el} * C_{s-el}$

- $C_p = a_t * t^k + a_d * d^k$

in which $k = 1$ in the linear case, and $k = 2$ in the quadratic case.

The performance data without the *S-Net* are given by Tables T1, and T2 (time and distance, respectively), and performance with *S-Net* is given by Tables T3 and T4. We compare the two systems (without and with *S-Net*) by computing the number of cases for which the *S-Net* system outperforms the non-S-Net system (over the 100 cases of experiment).

The time used for system without *S-Net* is:

trips robots	1	2	3	4	5	6	7	8	9	10
1	12.29	13.82	15.30	16.68	18.06	19.43	20.77	41.25	56.67	69.01
2	24.45	25.98	27.43	28.83	30.21	31.60	33.53	52.23	51.88	65.00
3	36.61	38.20	39.67	41.08	42.49	44.05	45.75	47.63	62.92	75.14
4	48.76	50.32	51.90	53.31	54.73	56.20	58.08	60.18	74.19	85.06
5	60.90	62.53	64.05	65.58	67.06	68.65	70.38	72.64	85.03	95.15
6	73.04	74.68	76.23	77.70	79.09	80.82	83.04	85.31	96.28	108.7
7	85.20	86.81	88.35	89.89	91.34	93.11	95.30	97.78	107.1	115.3
8	97.36	99.0	100.6	102.1	103.5	105.4	107.9	110.3	118.3	125.3
9	109.52	111.2	112.7	114.3	115.7	117.6	120.2	122.3	129.5	135.0
10	121.68	123.3	124.9	126.4	127.9	130.1	132.2	134.7	140.6	145.4

The distance traveled for system without *S-Net* is:

trips robots	1	2	3	4	5	6	7	8	9	10
1	122.7	124.6	126.7	128.5	130.4	132.2	133.8	232.1	305.3	362.8
2	244.1	246.0	247.9	249.8	251.6	253.4	257.1	346.3	339.0	400.1
3	365.5	367.6	369.7	371.5	373.5	375.7	378.9	382.6	454.3	510.6
4	486.8	488.8	491.0	492.9	494.8	497.2	500.8	505.5	570.5	620.0
5	608.0	610.4	612.6	614.7	616.9	619.6	622.6	627.9	684.8	730.4
6	729.2	731.7	733.9	735.8	737.7	741.0	745.8	751.2	800.9	857.9
7	850.6	853.0	855.2	857.3	859.4	862.8	867.4	873.5	915.1	950.5
8	972.0	974.5	976.7	978.7	980.7	984.9	990.2	996.2	1031	1060
9	1093	1096	1098	1100	1102	1106	1112	1117	1147	1169
10	1215	1217	1219	1222	1224	1229	1233	1239	1263	1280

The time used for system with *S-Net* is:

trips robots	1	2	3	4	5	6	7	8	9	10
1	12.18	13.18	14.18	15.23	16.26	17.25	18.26	19.25	20.24	21.24
2	24.09	25.17	26.17	27.29	28.31	29.27	30.38	31.42	32.52	33.51
3	35.99	37.16	38.17	39.28	40.28	41.35	43.91	44.90	45.92	46.94
4	47.93	49.20	50.17	51.28	52.30	53.43	55.87	56.99	57.99	59.16
5	59.83	61.18	62.11	63.28	64.36	65.41	69.14	70.11	71.02	71.27
6	71.75	73.20	74.08	75.39	76.41	77.60	80.99	82.03	82.96	84.14
7	90.43	91.87	92.63	93.95	96.28	94.06	95.43	95.53	95.03	96.81
8	101.3	102.8	103.6	104.9	107.1	106.4	106.8	107.2	107.5	108.8
9	112.15	113.8	114.5	115.8	117.9	117.6	118.2	119.0	119.8	121.1
10	123.0	124.8	125.4	126.9	128.8	129.4	129.5	131.1	131.9	133.0

The distance traveled for system with *S-Net* is:

trips robots	1	2	3	4	5	6	7	8	9	10
1	116.8	116.7	116.7	116.7	116.7	116.7	116.7	116.7	116.7	116.7
2	234.0	234.0	234.0	234.0	234.0	234.0	233.9	234.0	234.0	234.0
3	351.2	351.2	351.2	351.2	351.3	351.2	365.5	364.0	362.9	361.9
4	468.8	468.8	468.8	468.8	468.8	468.8	481.6	480.4	479.3	478.4
5	585.9	586.0	586.0	585.9	586.0	586.0	610.5	607.9	604.4	594.9
6	703.4	703.6	703.7	703.6	703.7	703.6	725.0	722.9	719.7	718.6
7	888.3	887.7	886.9	885.9	899.4	863.6	865.6	853.0	835.1	840.4
8	995.4	994.7	994.2	993.2	1005	983.1	975.9	965.3	956.2	954.8
9	1102	1102	1101	1100	1110	1092	1086	1078	1075	1073
10	1209	1209	1208	1207	1215	1207	1196	1194	1192	1187

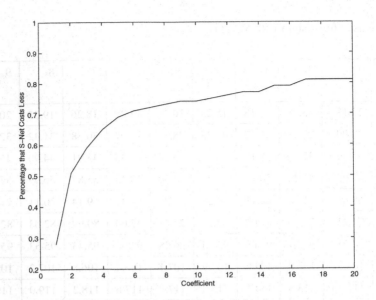

Figure 7.41: System Cost Comparison vs. Coefficient in Quadratic Distribution (adapted from [20]).

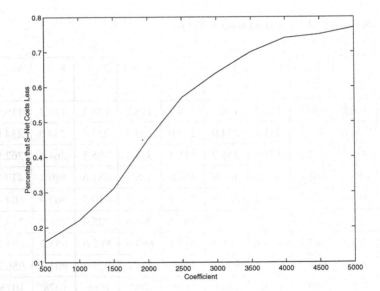

Figure 7.42: System Cost Comparison vs. Coefficient in Linear Distribution (adapted from [20]).

To establish C_s, we investigated mobile robot costs and a reasonable projection for *SEL* costs. For a given number of robots and *SEL*s, these costs are fixed and the cost variation comes from the C_p term. Rather than look at particular fixed a_t and a_d, we have assumed they are equal. Separate them out vary a_t independent of a_d. Figures 7.41 and 7.42 show the percentages of times the *S-Net* out-perform the *non-S-Net* as a function of the coefficient value ($a_t = a_d$). As can be seen, for both the quadratic and linear cost function, there are thresholds below which the *non-S-Net* out-performs the *S-Net*. This indicates that for any particular implementation, a specific detailed analysis should be done to determine which is preferred.

Chapter 8

CSN: The Heat Equation

We have proposed *Computational Sensor Networks* as a methodology[1] to exploit models of physical phenomena in order to better understand the structure of the sensor network. To do so, it is necessary to relate changes in the sensed variables (e.g., temperature) to the aspect of interest in the sensor network (e.g., sensor node position, sensor bias, etc.), and to develop a computational method for its solution. As examples, we describe the use of the heat equation to solve the sensor node localization problem and to detect sensor bias. Simulation and physical experiments are described.

A model-based approach to the design and implementation of *Computational Sensor Networks* (CSNs) is proposed. This high-level paradigm for the development and application of sensor device networks provides a strong scientific computing foundation, as well as the basis for robust software engineering practice. As described in Chapter 1, the three major components of this approach include (1) models of phenomena to be monitored, (2) models of sensors and actuators, and (3) models of the sensor network computation. We propose guiding principles to identify the state or structure of the phenomenon being sensed, or of the sensor network itself. This is called *computational modeling*. These methods are then incorporated into the operational system of the sensor network and adapted to system performance requirements to produce a mapping of the computation onto the system architecture. This is called *real-time computational mapping* and allows modification of system parameters according to real-time performance measures. This chapter deals mainly with computational modeling.

CSNs represent a scientific computing approach, and this includes the Verification and Validation (V & V) methodology of that discipline[118]; that is, model implementations must be verified (e.g., for correctness or numerical properties like error and convergence), and appropriate tests embedded in the system to monitor system correctness during execution. However, an important new aspect of this approach is that a CSN has the ability to sense and interact with the environment, and thus can run its own validation experiments to confirm or refute model structure or parameter values. Another intrinsic capability offered by CSNs is that models can be used to determine

---

[1]This chapter is a modified version of work done with Christopher Sikorski, Kyle Luthy and Edward Grant[66], and Felix Sawo and Uwe Hanebeck [132].

T.C. Henderson, *Computational Sensor Networks*, DOI: 10.1007/978-0-387-09643-8_8,
© Springer Science+Business Media, LLC 2009

unknown aspects of the structure of the measurement system itself given a known state of the physical phenomenon. For example, given the heat flow PDE and known temperatures at fixed (but unknown) sensor node locations, the equations can be reworked so as to determine the sensor locations (i.e., to solve the sensor localization problem). This can be done for a wide variety of initial conditions and depends only on the equations defining the physical process and the specific realization of the process in the world. Thus, real-time V & V are performed and this permits a scientifically repeatable basis for sensor network experiments. Real-time computational steering is achieved by (1) embedding verification and validation modules into the executable code, and (2) modeling module performance in terms of statistically meaningful characterization of output features conceptually defined by the user.

On the sensor network side, many advances have been made in sensor network technology and algorithms in the last few years. See [168] for an overview of the state of the art. Work has been done on: architecture [69], systems and security [167], and applications. Our own work has focused on the creation of an information field useful to mobile agents, human or machine, that accomplish tasks based on the information provided by the sensor network (see previous chapters). In order to address sensor networks in a comprehensive manner, the sensor network community has initiated a research program that includes work in the areas of sensor network architectures, programming systems, reference implementations, hardware and software platforms, testbeds and applications. Here we **explore the impact of a computational science approach on all these aspects of sensor networks**, and show that much benefit can be derived [56, 57]; in particular, the tools developed here can be highly leveraged across many scientific communities. CSNs will provide software engineering support for scientists and engineers to exploit sensor networks where it is notoriously difficult to develop and validate systems, for example, in our proposed snow monitoring application.

One of the major innovations of this approach is the incorporation of a strong model of the phenomenon to be observed. This allows the system developer great insight into the V & V requirements. We demonstrate the usefulness of the CSN approach by way of two examples:

- **Sensor Node Localization**: Given a model of the physical phenomenon, and a set of sensor nodes in unknown, but fixed, locations, use the computational model to determine the sensor node locations.

- **Sensor Bias Detection**: Given a model of the sensor, use the computational model to determine sensor bias.

8.1 Sensor Node Localization

To demonstrate how this methodology can be applied, we show how the sensor node localization problem can be solved. Oftentimes sensor devices are dropped at random into an environment or maybe moved (e.g., in a snow monitoring application, the devices may move with the snow both in depth as well as tangential to the surface). Many approaches to sensor node localization have been proposed [23, 102, 114]; see [160] for

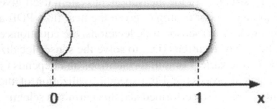

Figure 8.1: Heat Flow in a Uniform Rod (adapted from [66]).

a survey. As one example, Whitehouse and Culler propose a macro-calibration method for localization [162]. Their ad hoc localization system estimates distance between nodes using received signal strength information and acoustic time of flight. Although these phenomena can be modeled in the CSN context, their approach requires additional sensors (microphones) and processes. Moreover, CSNs solve an inverse problem based on the physical phenomenon - the example given here uses the heat equation (note that temperature sensors are ubiquitous and the method is robust).

Consider a rod of uniform cross-section and length 1 that is completely isolated except at the ends (see Figure 8.1). The heat flow is therefore limited to the x direction and the development of the temperature y over time can be described by the following partial differential equation (known as the *diffusion equation*):

$$\frac{\partial y}{\partial t} = D \cdot \frac{\partial^2 y}{\partial x^2} \ with \ D = \frac{\kappa}{c \cdot \rho}$$

where κ denotes the thermal conductivity, c the specific heat capacity and ρ the density of the rod. Figure 8.2 shows how the temperature changes over time for an arbitrary initial state. Note that usually the temperatures at the ends are fixed and the temperatures settle to a linear ramp (one could then easily assign locations to the nodes given a temperature); however, the basic requirement is that the temperature values in the rod change according to the heat equation in order for the method to work. It is also possible to allow the temperature at the ends to vary. There exist temperature distributions which are ambiguous, and thus where the method will not work – e.g., a constant temperature across the whole rod.

Such PDE's are usually solved by discretization and approximation of the derivatives. Then the temporal variation of the rod at any location can be determined using the standard finite difference approach: a grid of discrete, general points over the domain is considered and the derivatives are replaced by their finite-difference expressions at those points. We denote the points along the x-axis by x_i, the time points by t_j (with Δt the time step) and finally the temperature at point x_i and time t_j by $y_{i,j}$.

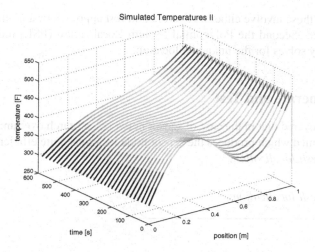

Figure 8.2: Simulation of Heat Flow Equation (adapted from [66]).

Then:

$$\left(\frac{\partial y}{\partial t}\right)_{i,j} = \frac{y_{i,j+1} - y_{i,j}}{\Delta t}$$

$$\left(\frac{\partial^2 y}{\partial x^2}\right)_{i,j} = \frac{\dfrac{y_{i+1,j} - y_{i,j}}{x_{i+1} - x_i} - \dfrac{y_{i,j} - y_{i-1,j}}{x_i - x_{i-1}}}{\frac{1}{2}(x_{i+1} - x_{i-1})}$$

which yields:

$$y_{i,j+1} = y_{i,j} + \frac{2\Delta t D}{(x_{i+1} - x_{i-1})}\left(\frac{y_{i+1,j}}{(x_{i+1} - x_i)}\right.$$
$$\left. - \frac{y_{i,j}}{(x_{i+1} - x_i)} - \frac{y_{i,j}}{(x_i - x_{i-1})} + \frac{y_{i-1,j}}{(x_i - x_{i-1})}\right)$$

The key idea is that the equations express an explicit relation between three positions on a line (two known endpoints and one unknown location between them), and four temperature values (all known and one at each location at time t and one at the unknown location at time $t + 1$). In general, this leads to a system of polynomial equations of degree three, however, for the case of one unknown location, this reduces to a single quadratic equation. This can the be solved and the root selected which best fits the conditions (e.g., must be between the two known locations).

To solve the localization problem in this case, the set of equations (one for each y_i) must be solved for the x_i values. This requires solving a set of degree 3 polynomial equations - which can be a difficult problem. For example, given n sensor nodes, there are up to 3^n distinct solutions (most are complex solutions, and thus not feasible). See [147] for analytical solution methods, e.g., homotopy continuations.

In the next sections we describe several alternative methods to solve for the *SEL* locations. First, a set of techniques are proposed which exploit the forward solution of the equation; these involve either a *generate and test* approach or a nonlinear gradient descent solver. Second the Polynomial System Localization (PSL) method is given which directly solves for the unknown location.

8.1.1 Generate and Test

We have discovered that in the case of sensor networks, a search over uniform samples can be performed which produces the sensor node locations quite efficiently. Consider *Algorithm Mesh_localization.*

Algorithm Mesh_localization

On input:

    n: the number of unknown *SEL* locations

    $T_i^{(j)}$: the temperature at node i at time j, $i = 1 \ldots n$

    x_0, T_0: min x value and temperature there

    x_{n+1}, T_{n+1}: max x value and temperature there

    δt: time step for simulation

    k: heat constant for simulation

On output:

    $S_i, i = 1 \ldots n$; *SEL* locations

begin

    $S_{ij} \leftarrow$ uniform mesh samples on interval

    $T'_{S_{ij}} \leftarrow Heat\_1D\_Sim$ – predicted temps for S_{ij}

    $D_{ij} \leftarrow \|T'_{S_{ij}} - T\|$ – distance from actual temps

    $i_{min}, j_{min} = argmin(D_{ij})$ – best temperature match

    $S_i \leftarrow S_{i_{min}, j_{min}}$

end

    The algorithm generates equi-spaced locations for the sensors, then simulates the heat equation to obtain predicted temperatures given the assumed node locations, then uses a distance norm to obtain an estimate of how much the actual and predicted temperature values differ. Finally, it determines the minimum error set of locations.

    The results of *Algorithm Mesh_localization* are good, but to achieve an answer in a reasonable time requires a hierarchical approach. First, locations are found at the integer level resolution for location, then these can be refined to get more accuracy. For example, for four unknown locations across an interval $[0, 11]$, each location has 10 possibilities for a total of 10^4 combinations (since the locations are not known to be in any particular order on the interval and cannot be at the endpoints of the interval).

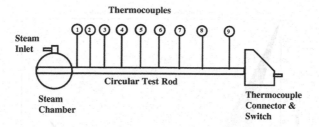

Figure 8.3: Layout of the Heated Rod Experiment (adapted from [66]).

8.1.2   Dense Sample Method

An alternate approach is to try to match more densely sampled temperature sets at an individual *SEL* to forward simulation temperature vs. time curves produced in the interval. *Algorithm Heat_1D_dense* does this.

Algorithm Heat_1D_dense

On input:

   n: the number of unknown *SEL* locations

   $T_i^{(j)}$: the temperature at node i at time j, $i = 1 \ldots n$

   x_0, T_0: min x value and temperature there

   x_{n+1}, T_{n+1}: max x value and temperature there

   δt: time step for simulation

   k: heat constant for simulation

On output:

   S_i, $i = 1 \ldots n$; *SEL* locations

begin

   $x_i \leftarrow$ dense set of samples on interval

   $T_i'^{(j)} \leftarrow Heat\_1D\_Sim$ – predicted temps at x_i at time j

   foreach T_k – temperature trace at unknown location

      $p_k = arg\{min\{correlation(T_k, T_i')\}\}$ – best temperature match

   end

   $S_i \leftarrow p_i, \quad i = 1 \ldots n$

end

   We have applied this method to data taken from an experimental apparatus (Figure 8.3 shows the layout). A one meter long stainless steel rod (304CG) of diameter one inch is connected to a steam chamber at one end and is instrumented with type T thermocouples located at 0.005m, 0.035m, and 0.095m, respectively, from the steam

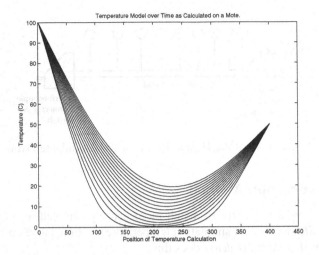

Figure 8.4: Forward Temperature Simulation from Tmote Sky Execution (adapted from [66]).

chamber. The thermocouples are connected to 10 channel selector switches which in turn are connected to a digital readout. The rod is attached to the steam chamber that provides a constant energy source at the base. The steam is turned on, and temperature readings are taken every 20 seconds as the rod heats.

Given knowledge of the initial conditions once the steam is activated (namely, 100 degrees C at one end and room temperature elsewhere along the rod), it is possible to run a careful simulation to obtain temperature curves at a dense sample of points along the rod (e.g., 1,000). Code was developed for the Tmote Sky and Figure 8.4 shows the results of a mote calculation. Each sensor is then matched independently to determine the best fit location.

Table 8.1 Simulated/Measured Temperature Data for Heated Rod.

$x = 0.005$ sim/measured	$x = 0.035$ sim/measured	$x = 0.095$ sim/measured
65.2/65.2	30.2/30.2	20.6/20.6
85.4/68.3	33.9/33.1	21.2/20.7
88.2/71.0	35.7/35.5	21.4/21.0
89.5/73.4	37.6/37.6	21.6/21.1
90.4/75.9	39.5/39.8	21.8/21.3
91.1/77.9	41.4/41.9	21.9/21.6
91.6/79.8	43.2/43.9	22.1/22.0

Table 8.1 gives the simulated and measured temperature values. The actual and recovered locations are: $(0.005, 0.035, 0.095)$ and $(0.011, 0.035, 0.100)$, respectively. As can be seen, the heat transfer model fits better away from the steam source.

8.1.3 Nonlinear Optimization Method

We have also studied the use of a nonlinear system solver (fminsearch[2] in Matlab) to solve the localization problem. This is set up as follows:

- Temperature values at specific locations and time instants are generated and taken as input data (the positions are not made available).

- A set of random location sets are generated, and the forward simulation used to predict the temperature at each time instant for each possible set of locations, and the lowest error set is used as the starting point for the nonlinear solution search. (Error is just the Euclidean distance between the temperature sets viewed as a vector.)

- Matlab's fminsearch is run to find the least error solution.

The results for one to four unknown *SEL* locations in the interval $[0, 10]$ and for four time instants ($t = 1, 2, 3, 4$ seconds) are shown in Table 8.2. Ten trials were run per test case, and of the 40 trials, 3 failed to find a solution.

Table 8.2 fminsearch Error Results.

Number Unknowns	Mean(max(errors))	Var(max(errors))
1	0.2281	0.1784
2	0.3529	1.2452
3	0.4016	0.3952
4	1.3395	10.3094

Thus, we see that for a small number of locations, and a reasonably good starting set of positions, the average error is low (less than 0.4 units out of 10 units), and has low variance. However, the method does not necessarily converge.

8.1.4 Polynomial System Localization (PSL)

The Polynomial System Localization (PSL method) is purely deterministic, meaning that neither uncertainties in the model description nor in the measurements are considered. This direct method is based on restating the model of the distributed phenomenon in terms of a polynomial system including the state of the phenomenon and the location to be identified. Then, solving a system of polynomial equations leads directly

---

[2]fminsearch performs multidimensional unconstrained nonlinear minimization using Nelder-Mead.

to the desired location of the sensor node. The PSL method has low computational complexity and can be implemented in a fairly straightforward manner.

This deterministic approach for the localization of individual nodes in a sensor network based on local measurements of a distributed phenomenon. The key idea of the proposed direct method is to solve the partial differential equation (Eqn 8.1) in terms of the unknown node locations. This leads to a straightforward solution as long as the resulting nonlinear equations can be readily solved. Solving these equations for all sensor locations is called the Polynomial System Localization Method. The PSL method basically consists of two steps: (1) spatial and temporal discretization of the mathematical model, and (2) reformulating and finally solving the resulting system of polynomial equations in terms of the desired locations.

Solving Polynomial System Equations

Based on the spatial and temporal discretization, the partial differential equation (Eqn 8.1) may be expressed as a finite difference equation and put in the following form at each discretization point, p_i, in the interval in question

$$0 = A_k^i(x_k^{i+1} - x_k^i)(x_k^i - x_k^{i-1})(x_k^{i+1} - x_k^{i-1}) - B_k^i(x_k^i - x_k^{i-1}) + C_k^i(x_k^{i+1} - x_k^i) \quad (8.1)$$

where

$$A_k^i = \frac{y_{k+1}^i - y_k^i}{2\alpha\Delta t}, \quad B_k^i = y_k^{i+1} - y_k^i, \quad C_k^i = y_k^i - y_k^{i-1}$$

At this point, it is important to mention that x_k^i represents the unknown location of the sensor node to be localized and x_k^{i+1} and x_k^{i-1} are the known locations of neighboring nodes. The derived system equation (Eqn 8.3) can be simply regarded as an explicit relation between three positions on a line (two known endpoints and one unknown location between them), and four values of the measured phenomenon (all known and one at each location at time t and one at the unknown location at time $t+1$). In order to derive the unknown location x_k^i of sensor node i, the polynomial system of equations (Eqn 8.3) needs to be solved and the root selected, which best fits the conditions (e.g., must be between the two known locations x_k^{i-1} and x_k^{i+1}).

The PSL method assumes a densely deployed sensor network in which every node i communicates with its neighboring nodes $i-1$ and $i+1$. This means that measurements of the distributed phenomenon y_k^{i-1} and y_k^{i+1} need to be transmitted between adjacent nodes. It can be stated that the denser the sensor nodes are deployed, the more accurate the individual nodes in the network can be localized. The proposed localization approach involves neither errors in the mathematical model nor uncertainties in the measurements. However, it can be easily implemented and has low computational complexity.

The simulation results for the PSL method are depicted in Figure 8.5. It is important to mention that this deterministic approach was simulated with perfect information, i.e., there is noise neither in the system nor in the measurements. Furthermore, we assume that the sensor node to be localized receives information about distributed phenomenon and locations from neighboring nodes. Since the diffusion equation has derivatives involving Δt and Δx, the PSL method is sensitive to the distance between the two

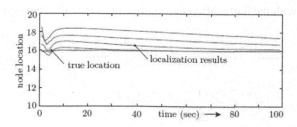

Figure 8.5: Results of PSL Method (adapted from [132]).

adjacent known locations. Evidence of this effect is shown in Figure 8.5 which plots the values found by the PSL method for known points of varying distance from the unknown. It is obvious that the denser the nodes are deployed the more accurate the location can be identified. The PSL method is a deterministic approach and is mainly based on restating the mathematical model in terms of the location. In the case of no noise in the model description and the measurement, this method leads to sufficient results for a dense sensor network.

In future work it is necessary to incorporate error into the PDE model as well as the sensors, and to study the robustness of the method in the presence of noise. Â Another issue is that if the locations of several nodes are unknown, they may be solved separately using the method described above; however, we should compare it to the simultaneous solution of the system of degree three equations. Finally, we intend to test the method on actual sensor data.

8.2   Sensor Bias Estimation

A sensor model generally characterizes at least two sources of error: (1) random error (*noise*), and (2) systematic error (*bias*). Given the exact temperature y_i^j at position x_i and time t_j, the measured value is given by:

$$z_i^j = y_i^j + b_i + \omega \qquad \omega \sim \mathcal{N}(0, \sigma^2)$$

where ω is random noise sampled from a Gaussian distribution with mean 0 and variance σ^2, and b_i is the bias of sensor S_i. Assume that $n + 2$ *SEL*s, and thus, the sensors, are equi-spaced along a rod of unit length with spacing h; that is, S_i is located at $x_i = ih$, $i = 0 \ldots n + 1$ (i.e., $x_0 = 0$ and $x_{n+1} = 1$). Thus, $h = \frac{1}{n+1}$.

Given the model of the heat equation and the sensor model, we will show how to determine values for the biases, $b_i, i = 1 \ldots n$. First, we rewrite (Eqn 8.1) as:

$$y_i^j = z_i^j - b_i - \omega$$

and combine it with the finite difference approximation to the heat equation:

$$\frac{y_i^{j+1} - y_i^j}{\Delta t} = D \cdot \frac{y_{i+1}^j - 2y_i^j + y_{i-1}^j}{h^2}$$

$$= D \cdot$$

$$\frac{(z_i^{j+1} - w_i^{j+1} - z_i^j + w_i^j)h^2}{\Delta t D}$$

$$-z_{i+1}^j + w_{i+1}^j + 2z_i^j - 2w_i^j - z_{i-1}^j + w_{i-1}^j = -b_{i+1} + 2b_i - b_{i-1}$$

Assume that the noise is small compared to the bias and note that the mean of the noise terms is 0; this yields:

$$-b_{i+1} + 2b_i - b_{i-1} = \frac{h^2(z_i^{j+1} - z_i^j)}{\Delta t D} - z_{i+1}^j + 2z_i^j - z_{i-1}^j$$

Now assume that the bias for S_0 and S_{n+1} is 0. Then this gives rise to a tridiagonal system:

$$\begin{pmatrix} -2 & 1 & & & \\ 1 & \ddots & & & \\ & \ddots & & 1 & \\ & & 1 & -2 \end{pmatrix} \cdot \begin{pmatrix} b_1 \\ \vdots \\ b_{n-1} \end{pmatrix} = \begin{pmatrix} v_1 \\ \vdots \\ v_n \end{pmatrix} \qquad \text{(v)}$$

where v_i is the right hand side of (Eqn v). This can be solved with direct methods (or in case of over-determined systems, can be solved using least squares).

We carried out some experiments to see how our approach for bias detection performs. We simulated temperatures using spectral methods to generate the data for the experiments. The sensor located at the center of the rod was assigned a bias of $b_{n/2} = 5$ while all other sensors had no bias. There are two sources of error to the bias estimate: inaccuracies introduced by the finite difference approximations and noise. Figure 8.6 demonstrates the influence of the former. The figure compares true and estimated bias values; $\Delta t = 2s$ was chosen as time step. For the estimation we used the approximation formula already introduced (error is $\in O((\Delta x)^2)$) as well as the following approximation formula which has error $\in O((\Delta x)^4)$:

$$\left(\frac{\partial^2 y}{\partial x^2}\right)_{i,j} = \frac{-y_{i-2,j} + 16y_{i-1,j} + 30y_{i,j} + 16y_{i+1,j} - y_{i+2,j}}{12(\Delta x)^2}$$

To apply this formula one has to assume that the two first and last bias values are known in advance (we set them to zero). As expected the higher accuracy of the latter formula also leads to better values for the b_i. The figure shows the influence of the sensors' distance Δx: For an increasing number of sensors (which decreases Δx) the error of the bias estimate is plotted. The error is computed by

$$e = \frac{1}{n+1}\left(\sum (b_i - \tilde{b}_i)^2\right)^{\frac{1}{2}}$$

where $\tilde{b}_i$ denotes the estimated value for b_i.

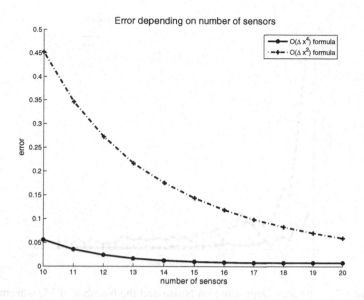

Figure 8.6: Errors Depending on Δx and the Approximation Formulas.

To demonstrate the influence of noise, we added white Gaussian noise to the (simulated) measurements. Now a single estimate is not sufficient any more since it is influenced by this noise. Therefore it becomes necessary to take into account more than just two consecutive measurements. This can be accomplished by using the equation several times for different measurements and solving this now over-determined system, or by solving the individual equations independently and computing the mean value. For different levels of noise Figure 8.7 shows how the bias estimate converges as more and more measurements are included (we used the same initial temperatures and bias values as above, 15 sensors and the $O((\Delta x)^4)$ finite-differences approximation).

So far we have done all of our sensor bias experiments with equispaced sensor nodes; this is a constraint that will rarely hold in real world applications and for unequally spaced sensors other approximations formulas have to be used, like derivatives of interpolating polynomials, for example (note that the resulting terms will still be linear in the b_i). The technique presented here also has another drawback. The computation is performed globally so that the measurements of all sensors first have to be sent to a central node before computation can start. But consider the following scenario: the sensor network could frequently test itself by computing the bias of a *single* sensor under the assumption that the neighboring sensors have zero bias (this computation can be done locally since it involves only measurements from a constant number of sensor-nodes – 3 sensors for the $O((\Delta x)^2)$-approximation and 5 sensors for the $O((\Delta x)^4)$-approximation). Although the condition of zero-biased neighboring nodes may not hold, the sensor-network should still be able to detect malfunctioning nodes this way and can exclude those from further computation.

We have demonstrated the ideas already presented for a concrete example: we estimated sensor bias solely based on sensor measurements and a system model by

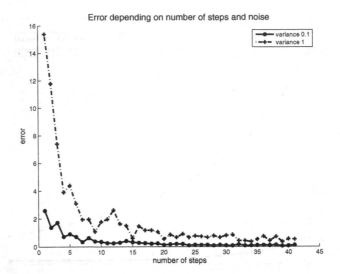

Figure 8.7: Estimates Depending on Noise and the Number of Measurements.

using methods known from the computational field. The accuracy of the approximation formulas had a clear impact on the reliability and the convergence behavior of the estimates – but more accurate approximations will also result in higher computation costs.

8.3 Future Directions

A major issue is the determination of an adequate model of the phenomenon. We take as our starting point that this is possible, especially when the structure of the model is known, and all that remains is to identify parameters. For recent work on this, see [135] and the next chapter. They derive the system model and the measurement model by the finite spectral method and show how nonlinear phenomena with complex boundary conditions can be reconstructed and predicted. More work needs to be done to characterize the types of functions which allow unique solutions in these circumstances.

These preliminary results are very encouraging. However, there is much work to be done:

- The method must be extended to other phenomena.

- The method must be extended to 2-D and 3-D.

- The method must be applied to physical experiments. This requires the identification of high-quality models (i.e, their form and parameters).

A few words are in order about the practicability of the method:

- Each sensor node can solve the problem independently in terms of sensor data from its neighbors.

- Generally, there will not be a large number of neighbors, and thus the system should be readily solvable.

- The only communication required is a time sequence sample of temperatures from the neighbors.

- Solutions can be shared between nodes to improve efficiency and accuracy.

- Temperature values can be averaged to reduce the effect of noise.

- Off-network computation of the numerical solution is also possible.

Acknowledgments
We would like to thank Mr. Konark Pakkala and Prof. Kent S. Udell for performing the heated rod experiment and collecting the data, Mr. Markus Westphal for his contributions to sensor bias estimation, Mr. Dietrich Brunn and Prof. Uwe Hanebeck for discussions on this topic, and Mr. Felix Sawo for his help and input.

Chapter 9

Bayesian Estimation of Distributed Phenomena

Felix Sawo and Uwe D. Hanebeck[1]

Intelligent Sensor-Actuator-Systems Laboratory
Institute of Computer Science and Engineering
Universität Karlsruhe (TH), Germany

This chapter introduces a Bayesian approach for the estimation of distributed phenomena based on discrete time-space measurements obtained by a sensor network. We introduce a new methodology for sensor network applications, which rigorously exploits mathematical models of the distributed phenomenon to be monitored. By this unobtrusive exploitation, the individual sensor nodes collect information not only about properties of the phenomenon but also about the sensor network itself. The novelty of the introduced estimation method is the systematic approach and the consideration of uncertainties not only occurring in the mathematical model but also arising naturally from noisy measurements.

First, it is shown how the physical phenomenon in terms of a *distributed*-parameter system description is spatially decomposed and temporally discretized leading to a *lumped*-parameter *finite*-dimensional description in state space form. Then, based on such a system description, the proposed methodology of *simultaneous state and parameter estimation* of distributed phenomena is introduced in a quite general form. It turns out that this in most cases leads to a *high-dimensional nonlinear* estimation problem, making special types of nonlinear estimators necessary. Accordingly, a novel estimator, the so-called *Sliced Gaussian Mixture Filter* is employed. This estimator exploits the linear substructure in the high-dimensional nonlinear estimation problem, and leads to a more efficient process. Furthermore, we introduce the application of this

---

[1]This chapter is a modified version of [132]

T.C. Henderson, *Computational Sensor Networks*, DOI: 10.1007/978-0-387-09643-8_9,
© Springer Science+Business Media, LLC 2009

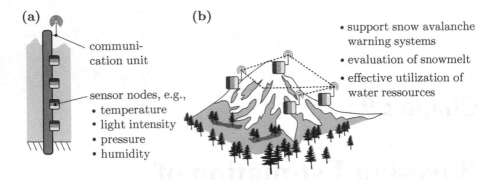

Figure 9.1: Visualization of a Snow Monitoring Scenario. **(a)** The individual sensor nodes collect local information about the snow state, e.g., temperature, light intensity, pressure, or humidity. **(b)** The observations provide useful information for snow avalanche warning systems and allow an effective utilization of water resources.(adapted from [132])

methodology to two of the most common tasks arising in sensor network applications. This results in two different methods:

- the *simultaneous reconstruction* of the state and *identification* of parameters of distributed phenomena (*SRI method*), and

- the *simultaneous reconstruction* of the state of distributed phenomena and *localization* of sensor nodes (*SRL method*).

The proposed methods provide novel prospects not only for the estimation of distributed phenomena but also for sensor network applications in general. Thanks to the *simultaneous approach*, the network is able to estimate the *entire* state of the distributed phenomenon, identify *non-measurable* quantities, verify and validate the correctness of the estimation results, and eventually adapt its algorithms and behavior in an autonomous fashion.

The results presented in this chapter were published in [14, 132, 134, 136]. However, the proposed model–based methods for the estimation of distributed phenomena are presented in a considerably extended form in this chapter.

9.1   Sensor Networks for Distributed Phenomena

In recent years, advances in technology have made it possible to build wireless sensor networks providing a smart interaction with the environment [25]. Typical advantages of using sensor networks include the deployment at low cost and in large numbers, as well as the inherent robustness thanks to the redundancy [21]. An important application for such networks is the observation of natural physical phenomena. Examples for such physical phenomena are: temperature distribution [136], chemical concentration [169], fluid flow, deflection and vibration in buildings, or the surface motion of a beating heart in minimally invasive surgery [5].

For the reconstruction of such distributed phenomena, the sensor network can be exploited as a huge information field collecting data from its surrounding. In such scenarios, the individual sensor nodes are densely deployed either inside the phenomenon or close to it. Then, by distributing local information to a global processing node, the phenomenon can be coöperatively reconstructed in an intelligent and autonomous manner [75, 130, 133]. This provides useful information both to mobile agents and to humans, which can accomplish their respective tasks more efficiently, thanks to the extended perception provided by the sensor network. Hence, dangerous situations, such as forest fires, seismic sea waves, or snow avalanches can be forecast or even prevented [66]. In the following, prospective application scenarios where sensor networks could provide a novel approach are described.

9.1.1 Prospective Application Scenarios

For *snow monitoring scenarios*, for example, there are two applications where the sensor network could provide novel possibilities: forecasting snow avalanches and flood runoffs. *Snow avalanches* are a major hazard to people, equipment or facilities, such as buildings, ski slopes, roads, power lines, and railways, in mountainous regions throughout the world. Each year, snow avalanches cause casualties and damage, not only in non-protected areas but also in popular cross-country skiing areas, e.g., in the Wasatch mountains in Utah. The application of an intelligent and autonomous sensor network could offer useful information for the support of avalanche forecasting systems. The individual sensor nodes deployed on the ground or within the snow pack collect measurable information about the snow state, such as temperature, light intensity, pressure or humidity; see Figure 9.1 **(a)**. Then, based on these observations and after further processing, measures about the stability of the snow pack, e.g., stress distribution, strain distribution, density distribution or location of so-called weak layers, of a certain area could be estimated [9, 99, 100, 127, 151]. Thus, by means of a sensor network, the possibility of snow avalanches can be predicted and defense structures in avalanche starting zones can be optimized; see Figure 9.1 **(b)**. An additional application scenario where sensor networks could provide novel possibilities is the accurate and efficient *evaluation of snowmelt*. By this means, water resources could be utilized more efficiently and flood runoffs could be forecast more accurately [76, 95].

A further example worth mentioning is the application of sensor networks to *monitoring* the condition and composition of ice in *skating rinks* [171]. For speed skaters to reach faster times, the optimal ice composition and especially the optimal temperature distribution of the ice is quite essential. For that reason, temperature nodes deployed at different points within the ice allow the estimation of the actual temperature distribution on the top of the surface and eventually the determination of the optimal ice composition. In addition, the sensor nodes can be linked to ice making machines, so that they can be adjusted in order to compensate changes in temperature, wind, or humidity [171].

In the aforementioned scenarios and for sensor network applications in general, the number of nodes and the measurement rates should be as low as possible due to economic and energetic reasons. As it stands, the lower the measurement rate of the individual nodes, the higher their durability. Therefore, a trade-off between energy costs and accuracy has to be found. The challenge for the observation of distributed phenomena is that measurements are available only at discrete time steps and discrete locations, meaning that no information between the individual sensor nodes is available. In order to get meaningful and accurate information not only at the sensor nodes itself but also between these nodes, the *model-based reconstruction* of the distributed phenomenon is of major significance. By exploiting additional physical background information of the phenomenon in the form of a mathematical model, the accuracy of the reconstruction can be improved significantly, even at non-measurement points [5, 130, 136].

In the following subsections, two of the *most important tasks* for the reconstruction of *distributed phenomena* based on a sensor network are explicitly described:

- identification of model parameters (SRI method, introduced in [134]), and

- localization of individual sensor nodes (SRLmethod, introduced in [132])

These two phases for the estimation of a distributed phenomenon are visualized in Figure 9.2. The novelty of the proposed methods is the rigorous exploitation of a mathematical model describing the dynamic and distributed behavior of both the phenomenon to be observed and the sensor network.

9.1.2   Parameter Identification (SRI method)

The model-based reconstruction of a distributed phenomenon by means of a sensor network is based on the mathematical model describing the physical behavior. Assuming we have an appropriate and sufficiently accurate model, the distributed phenomenon is uniquely characterized by model parameters and boundary conditions. However, in practical implementations, the model parameters such as the diffusion coefficient, might not be known in advance and usually need to be identified. Hence, one of the most important issues concerning distributed phenomena is the *parameter estimation*, also referred to as *parameter identification* or the *inverse problem*. The main goal is the estimation of parameters η_k^P in the system model from observed measurements such that the distributed state $p(z,t)$ sufficiently accurate explains the observations obtained by the sensor network [157]. The discrete time-space samples measured by the individual sensor nodes are incorporated into the mathematical model in order to improve its accuracy in terms of estimated model parameters [130].

For sensor network applications, the parameter identification becomes even more essential due to the harsh and unknown environment, unpredictable variations of the phenomenon, and imprecisely known sensor locations. It is important to emphasize that remaining uncertainties not only in the measurements but also in the assumed model structure need to be considered in a systematic way during the identification process. As it is shown in the following sections, the identification problem as well as the localization problem can be transformed into a *simultaneous state and parameter estimation* problem [133, 134]. Based on this framework, a Bayesian estimation approach can be employed, and thus the distributed phenomenon can be reconstructed and imprecisely known model parameters can be identified in a simultaneous fashion; see Figure 9.2 (a).

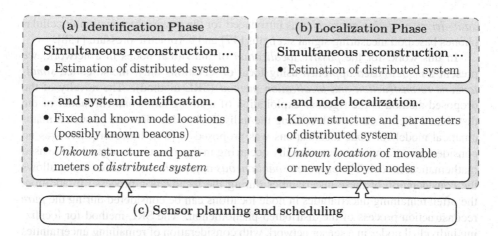

Figure 9.2: Visualization of *Two Phases* for the Estimation of Distributed Natural Phenomena. The phases are managed by a planning and scheduling process (not considered in this chapter). **(a)** The first phase consists of the identification of the environment in order to derive a mathematical model of the phenomenon to be monitored (*identification phase*). **(b)** Based on the mathematical model, newly deployed sensor nodes can be localized by local observations (*localization phase*)(adapted from [132]).

9.1.3 Node Localization (SRL method)

The sensor data derived from the individual nodes in most applications has only limited utility without location information. The precise knowledge of the node locations are particularly important for the accurate reconstruction of distributed phenomena. Manually measuring the location of every individual sensor node in the entire network becomes infeasible, especially when the number of sensor nodes is large or the nodes are inaccessible. The aforementioned issues make the localization problem one of the most important tasks to be considered in the area of sensor networks.

There are several ways to classify *localization methods*. In this research work, these methods are classified into *active methods* and *passive methods*. The *active localization methods* estimate the locations based on signals that are artificially stimulated and measured by the sensor network, e.g., artificially generated acoustic events. That means, the localization process is performed in fairly controlled environments, and incur significant installation and maintenance costs. A comprehensive survey on active localization methods can be found in [68].

In the case of *passive localization methods*, which rather occur in a non-controlled environment, the stimuli necessary for the localization process are generated in a natural fashion. The clear advantage of passive methods is that they do not need additional infrastructure. This certainly keeps the installation and maintenance costs at a very low level. In addition, these methods become particularly important for applications where satellite positioning systems are simply not available, e.g., sensor networks for monitoring the snow cover or indoor applications. In our previous research work, a purely *data–driven modeling approach* was introduced for the passive localization of cellular phones based on measuring signal strengths [47] and barometric pressure [158].

In this work, for the *passive localization* of individual nodes in a network, we present a *model-based approach* using local observations only, the so-called *simultaneous reconstruction and localization method* (SRL method). The novelty of the proposed method is the rigorous exploitation of a strong *mathematical model* of the distributed phenomenon for localizing the individual nodes. The use of such a mathematical model for node localizations was proposed in [66]; however, there was no consideration of uncertainties naturally occurring in the measurements and in the used mathematical model. The proposed *simultaneous approach*, on the other hand, allows the consideration of these uncertainties during the localization process. In addition, the often remaining uncertainties in node locations can be considered during the *pure* reconstruction process of the distributed phenomenon. The SRL method for localizing individual nodes in a sensor network with consideration of remaining uncertainties was introduced in [132].

It is shown that the localization problem can be regarded as a *simultaneous state and parameter estimation* problem, with node locations as the parameters to be identified. By this means, the sensor nodes are localized and the distributed phenomenon is reconstructed in a simultaneous fashion; see Figure 9.2 **(b)**. The improved knowledge can be exploited for other nodes to localize themselves.

9.2   Problem Formulation

There are several possibilities for the classification and characterization of physical phenomena and their respective mathematical descriptions. In this work, they are classified into *lumped*-parameter systems and *distributed*-parameter systems [14]. The key characteristic of a lumped-parameter system is that the state vector uniquely describing the system behavior depends only on time. Examples of lumped-parameter systems are bird flocks or swarm of robots. Such systems are usually described by a system of *ordinary* differential equations. On the other hand, the so-called *distributed state* of distributed-parameter systems does not only depend on time but also on the location, e.g., irrotational fluid flow, heat conduction, and wave propagation. The dynamic behavior of distributed-parameter systems can be described by a system of *partial* differential equations.

In this work, for simplicity we consider only distributed-parameter systems represented by *one-dimensional linear partial differential equations*, although similar expression can be found for the multi-dimensional case. In its most general form, the one-dimensional partial differential equation is given in implicit form by

$$\mathbb{L}\left(\boldsymbol{p}(r,t), \boldsymbol{s}(r,t), \frac{\partial \boldsymbol{p}}{\partial t}, \dots, \frac{\partial^i \boldsymbol{p}}{\partial t^i}, \frac{\partial \boldsymbol{p}}{\partial r}, \dots, \frac{\partial^i \boldsymbol{p}}{\partial r^i}\right) = 0, \qquad (9.1)$$

where $\boldsymbol{p}(r,t)$ denotes the state of the distributed system at time t and location z. The source term $\boldsymbol{s}(r,t)$, the state $\boldsymbol{p}(r,t)$, and its derivatives are related by means of a linear operator denoted by $\mathbb{L}(\cdot)$. The dynamic behavior of the distributed phenomenon strongly depends on specific physical parameters collected in the parameter vector

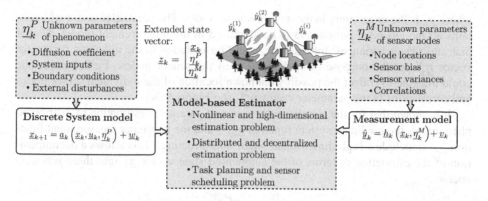

Figure 9.3: Overview and Challenges for the Model-based Simultaneous State and Parameter Estimation of Distributed Phenomena. Examples for unknown parameters $\underline{\eta}_k^P$ and $\underline{\eta}_k^M$ to be estimated in the system model and the measurement model(adapted from [132]).

$\underline{\eta}_k^P$, such as diffusion coefficient and coefficient of viscosity. The main goal is the estimation of the distributed state $\underline{p}(r, t)$ and the parameter vector $\underline{\eta}_k^P$ based on local measurements obtained by a sensor network.

In general, the application of a Bayesian estimation approach for the state and parameter estimation based on a distributed-parameter system (9.1) is a challenging task. For that reason, we presented in our previous research work [5, 133, 136] the conversion of the partial differential equation (9.1) into a *finite*-dimensional system in state-space form. The conversion of the system description leads to a high-dimensional nonlinear system model. This nonlinearity is mainly caused by the nonlinear relationship between the distributed state $\underline{p}(r, t)$ and unknown parameters $\underline{\eta}_k^P$. That means, the nonlinear finite-dimensional model of the distributed system (9.1) is given as follows

$$\underline{x}_{k+1} = \underline{a}_k\left(\underline{x}_k, \underline{\eta}_k^P, \hat{\underline{u}}_k\right) + \underline{w}_k^x, \tag{9.2}$$

where $\underline{x}_k$ represents the converted distributed state, $\hat{\underline{u}}_k$ denotes the system input, and $\underline{w}_k^x$ contains subsumed system uncertainties. The parameter vector $\underline{\eta}_k^P$ in (9.2) comprises all the *unknown parameters* to be identified in the distributed phenomenon, such as unpredictable variations of physical constants or material properties. In addition, unknown constraints at the boundary of the considered domain, unknown system inputs, and unknown disturbances could be included in the parameter vector $\underline{\eta}_k^P$; see Figure 9.3.

Besides the probabilistic system model there is a probabilistic measurement model describing the physical properties of the sensor network itself. That means, it relates the actual measurements of the network to the state vector $\underline{x}_k$ representing the distributed phenomenon. In this research work, we assume that the measurements $\hat{\underline{y}}_k$ are related *nonlinearly* to the state vector $\underline{x}_k$ according to

$$\hat{\underline{y}}_k = \underline{h}_k\left(\underline{x}_k, \underline{\eta}_k^M\right) + \underline{v}_k, \tag{9.3}$$

where $\underline{v}_k$ is the uncertainty in the measurement model. The parameter vector $\underline{\eta}_k^M$ contains all the *unknown parameters* to be identified in the measurement model. Sensor bias and sensor variances, for example, could be included in the unknown parameter vector $\underline{\eta}_k^M$ for the purpose of tracking wear of the sensor nodes. Furthermore, one could imagine to collect the possibly unknown location of the individual sensor nodes and correlations in the parameter vector $\underline{\eta}_k^M$; see Figure 9.3.

It is shown that for the simultaneous state and parameter estimation of distributed phenomena, the nonlinear system function $\underline{a}_k(\cdot)$ and the nonlinear measurement function $\underline{h}_k(\cdot)$ include a high-dimensional *linear* sub-structure. This allows a decomposition of the estimation in terms of the augmented state vector $\underline{z}_k$ into three sub-state vectors,

$$\underline{z}_k = \left[\underbrace{(\underline{x}_k)^T}_{\text{Linear subspace}} \quad \underbrace{(\underline{\eta}_k^P)^T \quad (\underline{\eta}_k^M)^T}_{\text{Nonlinear subspace}} \right]^T \tag{9.4}$$

with the high-dimensional state vector $\underline{x}_k \in \mathbb{R}^r$ (characterizing the conditional linear system) and the parameter vectors $\underline{\eta}_k^P \in \mathbb{R}^{N_p}$ and $\underline{\eta}_k^M \in \mathbb{R}^{N_m}$ (characterizing the nonlinear part of the system). For the estimation of the total state vector $\underline{z}_k$, the decomposition into a state vector $\underline{x}_k$ and parameter vector $\underline{\eta}_k$ is exploited for the derivation of a more efficient estimator than a nonlinear estimator operating on the entire vector $\underline{z}_k$. This decomposition of the estimation problem into a *linear* and a *nonlinear* problem is mainly achieved by a novel density representation, the so-called *sliced Gaussian mixture density*, and the systematic approximation of arbitrary densities by this representation.

9.3 Probabilistic Finite-Dimensional Models

The model–based state estimation of distributed phenomena based on a distributed–parameter description (9.1) is quite complex. The reason is that for a Bayesian estimation method usually a system description in lumped–parameter form is necessary. In order to cope with this, the distributed–parameter system is converted into a lumped–parameter system. Based on the resulting finite-dimensional model, algorithms can be derived allowing the estimation and identification of a distributed phenomenon (9.1), as well as the localization of individual sensor nodes locally measuring the distributed phenomenon.

In this section, we derive a finite-dimensional model of general distributed systems (9.1), which can be exploited for the simultaneous state and parameter estimation of such systems. The finite-dimensional model consists of two components: the system model and the measurement model. The *system model* describes the dynamic behavior of the distributed phenomenon to be monitored. The state of the phenomenon is uniquely characterized by a finite-dimensional state vector $\underline{x}_k$ and a vector $\underline{\eta}_k^P$ containing model parameters. On the other hand, the *measurement model* describes the distributed properties of the sensor network itself. The local measurements obtained by the individual nodes are related to both the state vector $\underline{x}_k$ and the parameter vector $\underline{\eta}_k^M$ containing for example node locations or sensor bias.

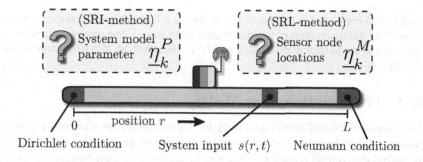

Figure 9.4: Visualization of the Solution Domain and Boundary Conditions of the Considered Distributed Phenomenon. The aim is the identification of system model parameters (SRI method) and the localization of sensor nodes (SRL method) based on local measurements of the phenomenon(adapted from [132]).

The methods introduced in this section can be applied to the general case of linear partial differential equations (9.1), and could even be extended to the multi-dimensional case in a straightforward fashion. However, we restrict our attention to a certain distributed phenomenon, the so-called diffusion equation.

Example 1 (Considered distributed phenomenon)
Throughout the entire chapter, we consider the following distributed phenomenon characterized by a one-dimensional partial differential equation

$$\mathbb{L}(\boldsymbol{p}(r,t)) = \frac{\partial \boldsymbol{p}(r,t)}{\partial t} - \alpha \frac{\partial^2 \boldsymbol{p}(r,t)}{\partial r^2} - \gamma s(r,t) = 0, \qquad (9.5)$$

where $\boldsymbol{p}(r,t)$ and $s(r,t)$ are the *distributed system state* and the *distributed system input*, respectively. The diffusion coefficient $\alpha \in \mathbb{R}$ is characterized by specific material properties, such as the medium density ρ, the heat capacity c_p, and the thermal conductivity k, according to $\alpha := \kappa/(\rho c_p)$. The input coefficient $\gamma \in \mathbb{R}$ represented by $\gamma := 1/(\rho c_p)$ characterizes the influence of the system input on the distributed phenomenon. For example, the propagation of heat in the snow pack can be described by such equations. The complete mathematical model of the snow pack, however, consists of further partial differential equations characterizing the model parameters α and γ; see [9, 99, 100]. These further dependencies of the parameters in terms of differential equations are omitted here for simplicity. For the derivation of a finite-dimensional model, and thus, the reconstruction of the entire distributed phenomenon (9.5), knowledge of the boundary conditions is necessary. There are several types of boundary conditions depending on the constraints at the boundaries of the considered solution domain. Considering the solution in a domain $\Omega = \{r | 0 \leq r \leq L\}$, we assume the following boundary conditions

$$\boldsymbol{p}(r = 0, t) = g_D, \qquad \frac{\partial \boldsymbol{p}(r = L, t)}{\partial r} = g_N, \qquad (9.6)$$

where g_N, specifying a condition on the derivative, is referred to as a *Neumann* boundary condition and g_D is the so-called Dirichlet boundary condition. The solution domain and the boundary conditions of the considered distributed phenomenon is visualized in

Fig 9.4. The main goal in this section is to derive a finite-dimensional model which can be used for the estimation of the distributed state $p(r, t)$, the model parameter α (SRI method) and the sensor node locations $\underline{\eta}^M(t)$ (SRL method) in a simultaneous fashion; see Fig 9.4.

9.3.1 Probabilistic System Model

The finite-dimensional model necessary for the simultaneous state and parameter estimation of distributed phenomena can be derived, in general, by methods for solving partial differential equations. The *modal analysis* method basically uses a set of *global* expansion functions for the approximation of the solution of the partial differential equation, and thus, the derivation of a probabilistic system model. These methods just need a few parameters for characterizing a smooth solution of the distributed phenomenon [128]. However, global expansion functions can be found only for simple problems with simple boundary conditions. On the other hand, there are methods such as the *finite-difference* method [22, 130], the *finite-element* method [6], and the *spectral-element* method [88, 94, 133]. The systematic decomposition of the solution domain involved in these methods allows the derivation of a probabilistic system model even for rather complex geometries and boundary conditions. Furthermore, the application to nonlinear partial differential equations is possible.

It is well-known that the aforementioned methods may be used with the same numerical methodology, the so-called *Galerkin formulation*. Based on this formulation, a finite-dimensional system model of the distributed phenomenon (9.5) can be derived in two steps, the *spatial decomposition* and the *temporal discretization*. The two steps for the conversion of the distributed phenomenon (9.5) and their respective resulting system description are visualized in Figure 9.5 **(a)**.

Spatial decomposition By means of the spatial decomposition, partial differential equations can be converted into a system of ordinary differential equations. First, the solution domain $\Omega = \{r | 0 \leq r \leq L\}$ is spatially decomposed into N_x subdomains Ω^e (the so-called *finite elements*). Then, an appropriate representation of the solution $p(r, t)$ within each subdomain Ω^e needs to be defined. The Galerkin method assumes that the solution $p(r, t)$ in the entire domain Ω can be represented by a piecewise approximation according to

$$p(r, t) = \sum_{i=1}^{N_x} \Psi_i(r) \, x_i(t), \qquad (9.7)$$

where $\Psi_i(r)$ are analytic functions called *shape functions* and $x_i(t)$ are their respective weighting coefficients. It is important to note that the individual shape functions $\Psi_i(r)$, in general, are defined in the entire solution domain Ω. The essence of the aforementioned finite-element and spectral-element method for the conversion of the distributed phenomenon lies in the choice of the shape functions $\Psi_i(r)$, e.g., piecewise linear functions, orthogonal functions, or trigonometric functions. The spatial decomposition into several subdomains Ω^e and the involving definition of respective shape functions $\Psi_i(r)$ are visualized by means of an example in Figure 9.5 **(b)**.

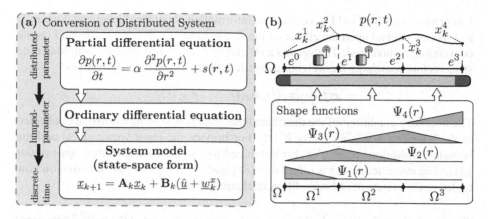

Figure 9.5: **(a)** Conversion of the Distributed System into a System Model in State-space Form (by spatial and temporal decomposition). **(b)** The solution $p(z,t)$ of the distributed phenomenon is approximated by a series of shape functions $\Psi_i(z)$ and their respective weighting coefficients x_k^i. Elemental decomposition of solution domain Ω into several subdomains Ω^e and application of shape functions $\Psi_i(z)$(adapted from [132]).

The approximated solution in terms of the finite expansion (9.7) cannot satisfy the partial differential equation (9.1) everywhere in the region of interest. That means usually a *residual* R_Ω remains. To make this residual small in some sense, a *weighted integral* has to be minimized

$$\int_\Omega \Psi_i(r)\,\mathbb{L}(p(r,t))\mathrm{d}r = 0,$$

with $i = 1,\ldots,N_x$. This weighted integral can be reduced to a system of *ordinary differential equations* by replacing the solution function $p(r,t)$ and the input function $s(r,t)$ by the finite expansion (9.7). In the case of the one-dimensional diffusion equation (9.5) this leads to following system of *ordinary* differential equations in terms of the continuous-time weighting coefficients $x_i(t)$,

$$\mathbf{M_G}\,\underline{\dot{x}}(t) = -\alpha\mathbf{D_G}\,\underline{x}(t) + \underbrace{(\gamma\,\mathbf{M_G}\underline{u}^*(t) + \underline{b}^*(t))}_{\underline{u}(t)}, \tag{9.8}$$

where $\mathbf{M_G}$ is called the *global mass matrix* and $\mathbf{D_G}$ is the *global diffusion matrix*. Basically, this system of equations describes the time evolution of the weighting coefficients $x_i(t)$ representing the approximated solution of the partial differential equation, i.e., approximation of the distributed state $p(r,t)$.

The individual entries M_{ij}^g and D_{ij}^g of the global mass matrix $\mathbf{M_G}$ and the global diffusion matrix $\mathbf{D_G}$ can be derived according to

$$M_{ij}^g = \int_\Omega \Psi_i(r)\Psi_j(r)\mathrm{d}r\,, \quad D_{ij}^g = \int_\Omega \frac{\mathrm{d}\Psi_i(r)}{\mathrm{d}r}\frac{\mathrm{d}\Psi_j(r)}{\mathrm{d}z}\mathrm{d}r.$$

It is obvious that $\mathbf{M}_G$ and $\mathbf{D}_G$ contain the information about the discretized domain Ω and merely depend upon the choice of the shape functions $\Psi_i(r)$, i.e., depend on the conversion method used. The vectors $\underline{x}(t)$ and $\underline{\dot{x}}(t)$ are the so-called continuous-time state vectors containing the weighting coefficients $x_i(t)$ and their derivatives

$$\underline{x}(t) = [\underline{x}_1(t), \underline{x}_2(t), \dots, \underline{x}_{N_x}(t)] \, .$$

The boundary conditions of the distributed phenomenon to be monitored are collected in the boundary condition vector $\underline{b}^*(t)$. For brevity, the input vector $\underline{u}^*(t)$ and the boundary condition vector $\underline{b}^*(t)$ are combined to a so-called *augmented input vector* $\underline{u}(t)$. The interested reader is referred to [88] and [136] for more information on how to derive the system of ordinary differential equations (9.8).

Temporal discretization   In the previous section, we presented the spatial decomposition allowing the conversion of the partial differential equation (9.5) into a set of ordinary differential equations (9.8). In this section, the time evolution of the distributed phenomenon can be specified by discretizing the system of ordinary differential equations (9.8) in time. The temporal discretization produces a linear system of equations in terms of the discrete-time state vector $\underline{x}_k$ containing the unknown weighting factors of the finite expansion (9.7).

To circumvent the restriction on the time step Δt, it is reasonable to integrate the set of ordinary differential equations by means of *implicit* methods, such as the Crank-Nicolson discretization. Basically, the Crank-Nicolson method evaluates the differential equation (9.8) at time step $t + \frac{1}{2}\Delta t$, approximates the time derivative on the left-hand side with a centered finite difference and the rest of the terms with averages. This approximation leads to following system of linear equations

$$\mathbf{M}_G \frac{\underline{x}_{k+1} - \underline{x}_k}{\Delta t} = \underline{u}_k - \frac{1}{2}\alpha \mathbf{D}_G \left[\underline{x}_{k+1} + \underline{x}_k \right], \tag{9.9}$$

where $\underline{x}_k$ denotes the spatially discretized state vector. It is important to note that this linear system is *unconditionally* stable for any selected time step Δt.

In the case of *linear* partial differential equations (9.5), the aforementioned methods for the spatial decomposition and temporal discretization result always in a *linear* system of equations according to

$$\underline{x}_{k+1} = \mathbf{A}_k(\alpha)\, \underline{x}_k + \mathbf{B}_k(\alpha)\, (\hat{\underline{u}}_k + \underline{w}_k^x) \, , \tag{9.10}$$

where $\underline{x}_k$ is referred to as the state vector characterizing the state of the distributed phenomenon. The system matrix $\mathbf{A}_k \in \mathbb{R}^{N_x \times N_x}$ and the input matrix $\mathbf{B}_k \in \mathbb{R}^{N_x \times N_x}$ are derived by

$$\mathbf{A}_k(\alpha) = \left(\mathbf{M}_G + \frac{1}{2}\alpha\Delta t \mathbf{D}_G \right)^{-1} \left(\mathbf{M}_G - \frac{1}{2}\alpha\Delta t \mathbf{D}_G \right),$$

$$\mathbf{B}_k(\alpha) = \Delta t\, \alpha \left(\mathbf{M}_G + \frac{1}{2}\alpha\Delta t \mathbf{D}_G \right)^{-1} \, .$$

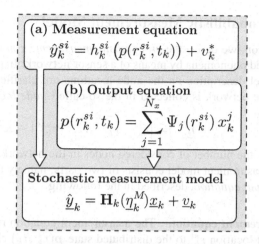

Figure 9.6: Components of the Probabilistic Measurement Model for the Estimation of Distributed Phenomena: **(a)** Measurement equation relating the measurements $\hat{y}_k^i$ to the distributed state $p(\underline{\eta}_k^M, t_k)$. **(b)** Output equation relating the distributed state $p(\underline{\eta}_k^M, t_k)$ to the finite-dimensional state vector $\underline{x}_k$ characterizing the phenomenon(adapted from [132]).

There are several important features to note about the finite-dimensional system model (9.10). It is obvious that the structure of the system matrix $\mathbf{A}_k$ and the input matrix $\mathbf{B}_k$ merely depend on the model parameters. In the case of the one-dimensional diffusion equation (9.5), the parameter vector $\underline{\eta}_k^P \in \mathbb{R}^{P_p}$ could contain the following model parameters

$$\underline{\eta}_k^P := \begin{bmatrix} \alpha & \gamma & \dots \end{bmatrix}^T \in \mathbb{R}^{P_p},$$

where α denotes the diffusion coefficient and γ is the system input coefficient. The parameters contained in the vector $\underline{\eta}_k^P$ are not restricted to the aforementioned parameters, but can be easily extended depending on the structure of the partial differential equation given in general form in (9.1).

That means, for the accurate reconstruction by means of a sensor network, parameters characterizing the behavior of the distributed phenomenon need to be precisely known. Due to such dependencies, the deviation of the true behavior from the probabilistic system model (9.10) leads to poor estimation results, shown by means of an example in Sec. 9.4. On the other hand, thanks to the dependency of the probabilistic system model (9.10) on such parameters, the identification problem can be stated as a *simultaneous state and parameter estimation* problem. Hence, the phenomenon can be reconstructed and unknown parameters can be identified in a simultaneous fashion.

Besides the finite-dimensional model of the distributed phenomenon, the mapping of specific measurements to the finite-dimensional state vector $\underline{x}_k$ representing the distributed state $p(r, t)$ is necessary. The probabilistic measurement model is introduced in the next section.

9.3.2 Probabilistic Measurement Model

In this section, we derive the *probabilistic measurement model* for the reconstruction of distributed phenomena by means of a sensor network. In particular, the dependency of the model description on the node locations are clarified. The node coordinates of the entire network is collected in the so-called *node location vector* $\underline{\eta}_k^M \in \mathbb{R}^M$, according to

$$\underline{\eta}_k^M := \begin{bmatrix} r_k^{s1} & r_k^{s2} & \cdots & r_k^{sM} \end{bmatrix} \in \mathbb{R}^M,$$

where M is the number of considered nodes in the network. For distributed phenomena, the measurement model consists of two parts, namely the *measurement equation* and the *output equation*, described in the following.

Measurement equation   The measurement equation relates the actual measurements $\hat{y}_k^i$ at location r_k^{si} to the distributed state $\boldsymbol{p}(r_k^{si}, t_k)$ characterizing the physical phenomenon, according to

$$\hat{y}_k^{si} = h_k^{si}\left(\boldsymbol{p}(r_k^{si}, t_k)\right) + \boldsymbol{v}_k^*,$$

where $\boldsymbol{v}_k^*$ contains the possibly correlated uncertainties arising from the sensor network. In general, depending on the measurement principle used for the actual sensor, the mapping $h_k^{si}(\cdot)$ consists of nonlinear functions; see Figure 9.6 **(a)**.

Output equation   The output equation relates the distributed state $\boldsymbol{p}(r_k^{si}, t_k)$ of the partial differential equation (9.5) in continuous space directly to the finite-dimensional state vector $\underline{x}_k$, according to

$$\boldsymbol{p}(r_k^{si}, t_k) = \sum_{j=1}^{N_x} \Psi_j(r_k^{si})\, \boldsymbol{x}_k^j,$$

where $\Psi_j(r)$ represents the *shape functions*. It is important to emphasize that the shape functions $\Psi_j(r)$ here are identical to the shape functions in the finite expansion (9.7) used for the *spatial decomposition*; see Figure 9.6 **(b)**.

Measurement model   By means of the measurement equation and the output equation, the entire measurement model for the estimation of distributed phenomena can be derived. For simplicity and brevity, we assume that the individual sensor nodes directly measure a realization of the distributed phenomenon $\boldsymbol{p}(r_k^{si}, t_k)$ at their respective locations r_k^{si}. Then, the measurement matrix $\mathbf{H}_k$ for the entire network is assembled by the individual shape functions,

$$\underline{\hat{y}}_k = \underbrace{\begin{bmatrix} \Psi_1(r_k^{s1}) & \cdots & \Psi_N(r_k^{s1}) \\ \vdots & \ddots & \vdots \\ \Psi_1(r_k^{sM}) & \cdots & \Psi_N(r_k^{sM}) \end{bmatrix}}_{\mathbf{H}_k(\underline{\eta}_k^M)} \underline{x}_k + \underline{v}_k, \qquad (9.11)$$

where $\underline{v}_k$ denotes the measurement uncertainty and M represents the number of sensor nodes used in the network. The measurement model (9.11) directly relates the measurements $\underline{\hat{y}}_k$ to the state vector $\underline{x}_k$ and to the location vector $\underline{\eta}_k^M$ containing the individual node locations. The components of the measurement model for the estimation of distributed phenomena is shown in Figure 9.6. In the following example, the structure of the measurement matrix $\mathbf{H}_k$ for the reconstruction and the localization is visualized.

Example 2 (Measurement model for node localization)

In this example, we clarify the structure of the measurement matrix $\mathbf{H}_k$ subject to *piecewise linear shape functions*. The entire solution domain Ω is decomposed into 3 subdomains and appropriate piecewise linear functions are defined on each sub-domain. The spatial decomposition and the shape functions are shown in Figure 9.5 **(b)**. Assuming there are two sensor nodes located at r_k^{s1} and r_k^{s2} in the sub-domains Ω^1 and Ω^2, the probabilistic measurement model is given as follows

$$
\begin{bmatrix} \hat{y}_k^1 \\ \hat{y}_k^2 \end{bmatrix} = \begin{bmatrix} \overbrace{c_1^1 + c_2^1\,r_k^{s1}}^{\Psi_1(r_k^{s1})} & \overbrace{c_3^1 + c_4^1\,r_k^{s1}}^{\Psi_2(r_k^{s1})} & 0 & 0 \\ 0 & \underbrace{c_1^2 + c_2^2\,r_k^{s2}}_{\Psi_2(r_k^{s2})} & \underbrace{c_3^2 + c_4^2\,r_k^{s2}}_{\Psi_3(r_k^{s2})} & 0 \end{bmatrix} \begin{bmatrix} x_k^1 \\ x_k^2 \\ x_k^3 \\ x_k^4 \end{bmatrix} + \underline{v}_k.
$$

The constants c_i^j arise from the definition of the piecewise linear shape functions in each sub-domain, i.e., the geometry of the applied grid for the finite elements. The extension to orthogonal polynomials and trigonometric functions can be derived in a straightforward fashion.

There are several important properties of the measurement model (9.11) essential for the estimation of distributed phenomena and the localization of sensor nodes based on local measurements. It is obvious that the structure of the measurement matrix $\mathbf{H}_k$ merely depends on the location $\underline{\eta}_k^M$ of the individual sensor nodes. That means, for the accurate reconstruction of the distributed phenomenon (9.5) based on a sensor network, the exact node locations $\underline{\eta}_k^M$ are necessary. Due to this dependency, deviations of true locations from the modeled node locations lead to poor estimation results. This degradation of the estimation performance is shown in Sec. 9.4.

On the other hand, thanks to the dependency of the measurement matrix $\mathbf{H}_k$ on the node locations $\underline{\eta}_k^M$, the localization problem can be stated as a *simultaneous state and parameter estimation* problem. By this means, the distributed phenomenon can be reconstructed and the sensor nodes can be localized in a simultaneous fashion. The method for the *simultaneous reconstruction and node localization* (SRL method) is introduced in Sec. 9.7.

9.4 Reconstruction of Distributed Phenomena

The probabilistic finite-dimensional model (9.10) can be used for the *simulation of the distributed phenomenon* by simply propagating the finite-dimensional state vector $\underline{x}_k$ over time. Based on this propagation, the distributed state $p(r, t)$ of the underlying

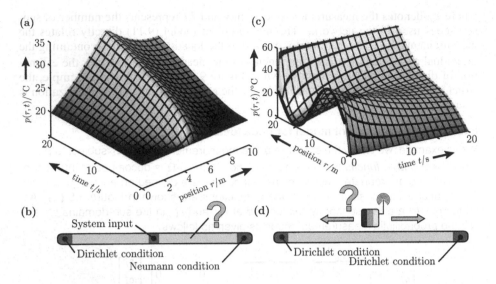

Figure 9.7: Visualization of the Numeric Solution **(a)**,**(c)** and their Respective Assumed Conditions **(b)**,**(d)** of the Simulated Distributed Phenomenon. These examples are used for demonstrating the performance of the proposed methods, i.e., for *pure reconstruction*, *identification* of model parameters (SRI method), and *localization* of sensor nodes (SRL method)(adapted from [132]).

phenomenon is directly derived using the finite expansion (9.7) for given initial conditions. However, for the model-based estimation of distributed phenomena, the aim is not just the simulation of the system, but the *reconstruction* of the entire state $p(r, t)$ by means of *measurements* obtained from a *sensor network*.

This section is devoted to the *state reconstruction* of the distributed phenomenon by means of discrete time-space measurements only, i.e., the mathematical model and the node locations are precisely known. The introduced reconstruction process allows to derive estimates not only at the actual measurement points but also at non-measurement points. It is shown that by assuming a precise mathematical model this process leads to accurate estimation results. On the other hand, the deviation of parameters such as diffusion coefficient $\underline{\eta}_k^P$ or node locations $\underline{\eta}_k^M$ leads to a degradation of the performance of the entire reconstruction process.

9.4.1   Reconstruction based on Precise Mathematical Models

In general, depending on the structure of the system model (9.2) and the measurement model (9.3), i.e., being linear or nonlinear, an appropriate estimator has to be chosen in order to estimate the state characterizing the distributed phenomenon. For the *pure reconstruction* of the distributed state $p(r, t)$ the system model (9.10) and the measurement model (9.11) are linear in terms of the state vector $\underline{x}_k$. Hence, it is sufficient to use the linear Kalman filter to obtain the best possible estimate, and eventually reconstruct the entire phenomenon. The reconstruction process of the distributed phenomenon based on the linear Kalman filter consists of two steps, the *linear prediction step* and the *linear measurement step*. These two steps are alternately performed in order to reconstruct the entire distributed state $p(r, t)$ even at non-measurement points, as visualized in Figure 9.8.

Linear Prediction Step   The purpose of the *linear* prediction step is to propagate the current state estimate $\underline{\hat{x}}_k^e$ through the *linear* system equation (9.2) to the next time step. In the case of the Kalman filter, the probabilistic of the general random vector $\underline{x}_k$ is uniquely characterized by the mean $\underline{\hat{x}}_k$ and the covariance matrix $\mathbf{C}_k$. For the pure reconstruction, we assume a precise mathematical model (9.2) of the underlying distributed phenomenon, i.e., a precisely known diffusion coefficient. Hence, the mean $\underline{\hat{x}}_{k+1}^p$ and the covariance matrix $\mathbf{C}_{k+1}^p$ of the state vector $\underline{x}_{k+1}$ can be simply calculated by

$$\underline{\hat{x}}_{k+1}^p = \mathbf{A}_k \underline{\hat{x}}_k^e + \mathbf{B}_k \underline{\hat{u}}_k,$$
$$\mathbf{C}_{k+1}^p = \mathbf{A}_k \mathbf{C}_k^e \mathbf{A}_k^T + \mathbf{B}_k \mathbf{C}_k^w \mathbf{B}_k^T, \tag{9.12}$$

where $\underline{\hat{x}}_k^e$ and $\mathbf{C}_k^e$ are the mean and the covariance matrix of the estimated state vector $\underline{x}_k$ from the previous time step. The covariance matrix $\mathbf{C}_k^w$ represents the input uncertainties. It is important to note that, for simplicity and brevity, we assume the input vector $\underline{u}_k$ and the state vector $\underline{x}_k$ to be stochastically uncorrelated.

Linear Measurement Step   For the purpose of reducing the uncertainty of the state vector $\underline{x}_k$, measurements $\underline{\hat{y}}_k$ obtained from the sensor network are incorporated into the reconstruction process. For distributed phenomena, the discrete time-space measurements $\underline{\hat{y}}_k$ are related to the state vector $\underline{x}_k$ via the measurement model (9.11) derived in the previous section. Assuming a precise measurement matrix $\mathbf{H}_k$, i.e., precisely known node locations and sensor characteristics, the mean $\underline{\hat{x}}_k^e$ and covariance matrix $\mathbf{C}_k^e$ of the estimated state $\underline{x}_k$ can be derived by

$$\underline{\hat{x}}_k^e = \underline{\hat{x}}_k^p + \mathbf{C}_k^p \mathbf{H}_k^T \left(\mathbf{C}_k^v + \mathbf{H}_k \mathbf{C}_k^p \mathbf{H}_k^T \right)^{-1} \left(\underline{\hat{y}}_k - \mathbf{H}_k \underline{\hat{x}}_k^p \right),$$
$$\mathbf{C}_k^e = \mathbf{C}_k^p - \mathbf{C}_k^p \mathbf{H}_k^T \left(\mathbf{C}_k^v + \mathbf{H}_k \mathbf{C}_k^p \mathbf{H}_k^T \right)^{-1} \mathbf{H}_k \mathbf{C}_k^p. \tag{9.13}$$

The matrix $\mathbf{C}_k^v$ denotes the possibly correlated covariance matrix of the individual nodes in the entire sensor network. For simplicity and brevity, we assume the measurements $\underline{y}_k$ to be stochastically uncorrelated to the state vector $\underline{x}_k$.

The performance of the reconstruction process assuming a precise mathematical description of the underlying distributed phenomenon and the sensor network itself is demonstrated by means of the following example.

Example 3 (Precise mathematical models)
In this example, the performance of the reconstruction method in the case of precise mathematical models is demonstrated by means of simulation results. The goal is the

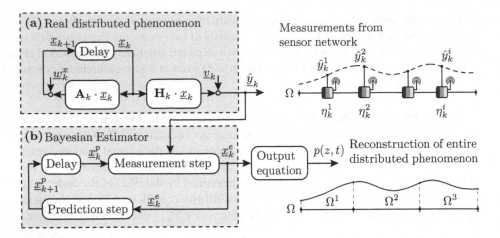

Figure 9.8: Structure of a Distributed Phenomenon Represented as a High-dimensional Linear System with Additive Noise and its Respective Linear Bayesian Estimator for the Reconstruction. The output $\hat{\underline{y}}_k$ can be regarded as a realization of the random variable $\underline{y}_k$. The input $\underline{u}_k$ is omitted here for simplicity. By means of the model-based estimation process the entire phenomenon can be reconstructed, even at non-measurement points(adapted from [132]).

reconstruction of the distributed state $p(z, t)$ using both a mathematical model describing the physical behavior and measurements obtained by a single sensor node. It is important to emphasize that the novelty of the proposed approach is to consider remaining uncertainties arising from noisy measurements and occurring in the mathematical models. We assume the underlying phenomenon to be represented by the one-dimensional partial differential equation (9.5), introduced in Example 1. The distributed-parameter system (9.5) is converted into a lumped-parameter system based on *piecewise linear* shape functions, i.e., using the finite element method. For simplicity, the sensor network consists of one single sensor node at location η_k^S.

For the *pure reconstruction* of the entire phenomenon using just a single sensor node, we assume the model parameter η_k^P and the node location η_k^S to be precisely known. The nominal parameters of the mathematical model of the phenomenon and the sensor node are given by:

$$
\begin{aligned}
\text{Solution domain} \quad & L = 10\,\text{m}, \\
\text{Dirichlet condition at left end} \quad & g_D = p(r = 0, t) = 20\,^\circ\text{C}, \\
\text{Neumann condition at right end} \quad & g_N = \frac{\partial p(r = L, t)}{\partial r} = 0\,\text{m}^{-1} \cdot {}^\circ\,\text{C}, \\
\text{Model parameter / node location} \quad & \alpha_k = 0.8\,\text{m}^2\,\text{s}^{-1},\ \gamma_k = 1\,\text{cal}^{-1} \cdot \text{m}^3 \cdot {}^\circ\,\text{C},\ r_k^s = 8\,\text{m}, \\
\text{Time discretization constant} \quad & \Delta t = 0.01\,\text{s}, \\
\text{Number of discretization nodes} \quad & N_x = 50, \\
\text{System input} \quad & s(r, t) = 10 \cdot e^{-10\,(r-5)^2}\,\text{cal} \cdot \text{m}^{-3} \cdot \text{s}^{-1}.
\end{aligned}
$$

The assumed conditions of the simulated example and the numeric solution of the deterministic partial differential equation for a given initial solution is depicted in

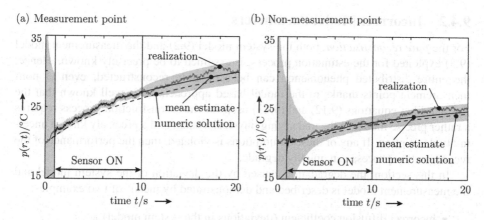

Figure 9.9: Realization of the Distributed Phenomenon $p(z, t)$ (gray), Mean of Reconstructed Phenomenon (black), 3σ-bounds (gray shaded), and Numeric Solution of Deterministic Model (black dotted) for **(a)** measurement point and **(b)** non-measurement point(adapted from [132]).

Figure 9.7 **(a)-(b)**. Based on the system model (9.2) and the measurement model (9.11) with aforementioned nominal parameters, the estimator for the purpose of reconstructing the distributed phenomenon can be designed. The noise terms represented by respective covariance matrices are assumed as follows

System input noise $\mathbf{C}_k^w = \mathrm{diag}\,\{100, \dots, 100\}\ \mathrm{cal}\cdot\mathrm{m}^{-3}\cdot\mathrm{s}^{-1}$,

Measurement noise $C_k^v = 1\,^\circ\mathrm{C}$.

The simulation results are depicted in Figure 9.9. It is obvious that using a model-based approach the entire distributed phenomenon can be reconstructed. At the beginning of the simulation just uncertain information about the distributed state $p(z, t)$ is known. As soon as the sensor node starts to measure a realization of the phenomenon at a certain location, the estimation becomes more and more certain, i.e., the uncertainty (gray shaded area) decreases. This is depicted in Figure 9.9 **(a)**. It is important to emphasize, that the uncertainty decreases not only at the measurement point, but also at non-measurement points, thanks to the model-based approach; see Figure 9.9 **(b)**. Furthermore, based on the estimated phenomenon in terms of a density function, optimal measurement sequences and locations can be found using sensor planning and scheduling algorithms.

For the *pure reconstruction* introduced in this section, the model parameters $\underline{\eta}_k^P$ and the node locations $\underline{\eta}_k^M$ are assumed to be precisely known. As already mentioned, the deviation of the assumed mathematical models from both the real distributed phenomenon and the real properties of the sensor network leads to a degradation of the estimation performance. This is demonstrated in the next section.

9.4.2   Incorrect Model Parameters

For the *pure reconstruction*, both the system model (9.2) and the measurement model (9.3) exploited for the estimation process, is assumed to be precisely known. Hence, the entire distributed phenomenon can be accurately reconstructed, even at non-measurement points thanks to the model-based approach. It is well known that the Kalman filter equations (9.12) and (9.13) used for the reconstruction process requires a rather precise model of the underlying physical system and a precisely known uncertainty description. If any of these assumptions is violated, then the performance of the reconstruction process can quickly degrade.

In this section, the degradation caused by the deviation of the system model and the measurement model is described and demonstrated by means of two examples:

- incorrect diffusion coefficient (deviations in the system model)

- incorrect node locations (deviations in the measurement model)

These two examples demonstrate the severe effect of assuming parameters both in the system model and the measurement model deviating from the true system. Furthermore, this degradation of the performance justifies the simultaneous approach for the *parameter identification* (SRI method) and the *node localization* (SRL method) during the reconstruction of distributed phenomena.

Incorrect Diffusion Coefficient   In many cases, the underlying real physical phenomenon deviates from the nominal mathematical model, basically caused by neglecting particular physical effects or external disturbances. Furthermore, the respective model parameters could vary over time without knowing the exact dynamic behavior of these variations. In addition, due to the distributed characteristic of the physical phenomenon, not only the states are distributed and inhomogeneous but also the parameters describing the dynamic behavior. Considering all these issues in the mathematical model quickly increases the complexity of the model description and the computational load. On the other hand, neglecting these physical effects leads to a deviation of the mathematical model and thus, causes a degradation of the reconstruction performance. That means, for practical applications a trade-off between accuracy and computational load needs to be found. The degradation leading to poor performance is illustrated in the next example.

Example 4 (Reconstruction with incorrect model parameters)
In this example, we consider a distributed phenomenon represented by the one-dimensional partial differential equation (9.5) with respective boundary conditions and system inputs. The nominal parameters for the system model (9.5) are given by

$$\text{Dirichlet condition at left end} \quad g_D = p(r = 0, t) = 20\,^{\circ}\text{C},$$

$$\text{Neumann condition at right end} \quad g_N = \frac{\partial p(r = L, t)}{\partial r} = 0\,\text{m}^{-1} \cdot {}^{\circ}\text{C},$$

$$\text{System input} \quad s(r, t) = 10 \cdot e^{-10\,(r-5)^2}\,\text{cal} \cdot \text{m}^{-3} \cdot \text{s}^{-1},$$

$$\text{True model parameter} \quad \alpha_{\text{true}} = 0.8\,\text{m}^2 \cdot \text{s}^{-1},$$

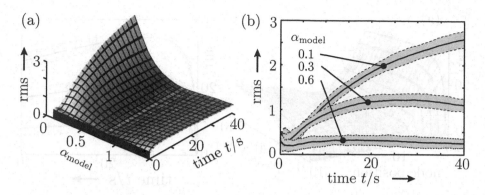

Figure 9.10: Root Mean Square Error $\hat{e}_k$ and Error Variance C_k^{rms} for the Estimated Distributed Phenomenon for 100 Monte Carlo simulation runs. The true parameter α_{true} is given by $\alpha_{\text{true}} = 0.8\,\text{m}^2 \cdot \text{s}^{-1}$. **(a)-(b)** Visualization of rms $\hat{e}_k$ of the Kalman filter based on various incorrect parameters $\alpha_{\text{model}} = \{0.1, 0.2, \ldots, 1.5\}\,\text{m}^2 \cdot \text{s}^{-1}$ (adapted from [132]).

where the remaining parameters necessary for the reconstruction are described in Example 3. The assumed boundary conditions, the location of the system input and the numeric solution of the deterministic partial differential equation (9.5) are visualized in Figure 9.7 **(a)-(b)**. The state estimation of the distributed phenomenon is performed on the basis of a Kalman filter with the nominal parameter set for the diffusion coefficient η_k^P according to

$$\alpha_{\text{model}} = \{0.1, 0.2, \ldots, 1.5\}\,\text{m}^2 \cdot \text{s}^{-1},$$

with the true parameter $\alpha_{\text{true}} = 0.8\,\text{m}^2 \cdot \text{s}^{-1}$. For each parameter value, 100 independent Monte Carlo simulation runs have been performed, resulting in $N_{\text{MC}} = 100$ true realizations $\underline{\tilde{x}}_k^i$ of the finite-dimensional state vector characterizing the distributed state $p(z,t)$. The simulation result for the reconstruction with incorrect model parameters is shown in Figure 9.10.

Based on the reconstruction process described in Sec. 9.4.1, the entire distributed phenomenon can be reconstructed using the nominal mathematical models and the discrete time-space measurements from the sensor network. The estimated finite-dimensional state vector $\underline{x}_k^e$ can be directly derived from (9.12) and (9.13). The root mean square error (rms) and the error variance for the estimation result are approximated by calculating the average according to

$$\hat{e}_k \approx \sqrt{\frac{1}{n \cdot m} \sum_{i=1}^{N_{\text{MC}}} \|\underline{\tilde{x}}_k^i - \underline{\hat{x}}_k^i\|}, \qquad C_k^{\mathrm{rms}} \approx \frac{1}{n-1} \sum_{i=1}^{n} \left(e_k^i - \hat{e}_k\right)^2, \qquad (9.14)$$

where $\underline{\hat{x}}_k^i$ denotes the mean of the estimated state vector $\underline{x}_k^e$. The root mean square error $\hat{e}_k$ and error variance C_k^{rms} for each nominal parameter value are visualized in Figure 9.10 **(a)-(b)**. It can be clearly seen that the more the nominal parameter α_{model} deviates from the true parameter α_{true}, the more the performance of the estimation results degrades.

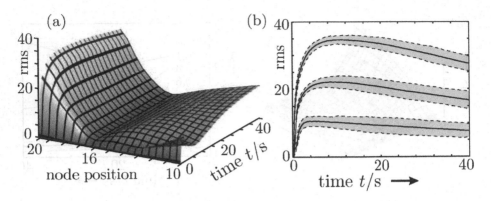

Figure 9.11: Visualization of Root Mean Square Error (rms) and Error Variance Averaged over 100 Monte Carlo Simulation Runs. The true node location is assumed to be $r^s_{\text{true}} = 16\,\text{m}$. **(a)** Rms of Kalman filter based on incorrect node locations $r^s_{\text{model}} = \{10, 10.5, \dots, 20\}\,\text{m}$. It is obvious that with the deviation of the node location the performance quickly degrades. **(b)** Comparison of *Kalman filter* based on incorrect node locations and the simultaneous reconstruction and node localization (*SRL method*) approach(adapted from [132]).

Incorrect Node Locations In many real world applications the actual properties of the sensor network deviate from the measurement model. This deviation of the mathematical model could be caused for example by deviated node locations, ignored sensor bias, or imprecisely known correlations between the nodes. In particular, the locations of sensor nodes (randomly deployed or movable nodes) contain some uncertainties or even could be *completely unknown*. The degradation of the performance of the reconstruction process caused by *deviated node locations* is demonstrated in the following example.

Example 5 (Reconstruction with incorrect node location)

In this example, we consider the one-dimensional diffusion equation (9.5) subject to Dirichlet boundary condition at both ends and respective initial conditions. The nominal parameters for the system model (9.5) and the measurement model are given by

Dirichlet condition at left end	$g^L_D = p(r = 0, t) = 0\,^\circ\text{C},$
Dirichlet condition at right end	$g^R_D = p(r = L, t) = 60\,^\circ\text{C},$
System input	$s(z, t) = 0\,\text{cal} \cdot \text{m}^{-3} \cdot \text{s}^{-1},$
True node location	$\eta^S_{\text{true}} = 16\,\text{m},$

where the remaining parameters are described in Example 3. The assumed boundary conditions and the numeric solution of the deterministic partial differential equation (9.5) are visualized in Figure 9.7 **(c)-(d)**. The system uncertainty at the individual discretization nodes is given by $C^{rwi}_k = 20$ and the measurement noise variance by $C^v_k = 0.5\,^\circ\text{C}$. The reconstruction of the distributed phenomenon is performed on the basis of a Kalman filter with nominal parameter set for the sensor location r^s_{model}

according to
$$r_{\text{model}}^s = \{10, 10.5, \ldots, 20\} \text{ m}.$$
For each assumed node location, 100 independent Monte Carlo simulation runs have been performed, resulting in $N_{\text{MC}} = 100$ true realizations $\tilde{x}_k^i$ of the distributed phenomenon.

The root mean square error (9.14) of the Kalman filter based on incorrect node locations is shown in Figure 9.11 **(a)**. It is obvious that the more the assumed node location η_{model}^S deviates from the true location η_{true}^S, the more the performance of the reconstruction result degrades. Figure 9.11 **(b)** depicts the comparison of the estimation error between the Kalman filter based on incorrect node locations and the simultaneous node localization and reconstruction approach (*SRL method*). Obviously, thanks to the simultaneous localization approach the performance of the reconstruction can be significantly increased.

9.5   Augmented Model for Node Localization

For the simultaneous state and parameter estimation of distributed phenomena, the unknown parameters $\underline{\eta}_k^P$ and $\underline{\eta}_k^M$ are treated as additional state variables. By this means, conventional estimation techniques can be used to *simultaneously* estimate the parameters, such as model parameters or node locations, and the state of the distributed phenomenon. Hence, an *augmented state vector* $\underline{z}_k$ containing the system state $\underline{x}_k$ and the additional unknown parameters is defined by (9.4).

The augmentation of the state vector with additional unknown parameters leads to the so-called *augmented finite-dimensional model* of distributed phenomena. In the case of the sensor node localization by local measurements (SRL method), the augmentation results in the following augmented system model and augmented measurement model

$$\begin{bmatrix} \underline{x}_{k+1} \\ \underline{\eta}_{k+1}^M \end{bmatrix} = \begin{bmatrix} \mathbf{A}_k\, \underline{x}_k + \mathbf{B}_k\, \hat{\underline{u}}_k \\ \underline{a}_k^M(\underline{\eta}_k^M) \end{bmatrix} + \begin{bmatrix} \mathbf{B}_k\, \underline{w}_k^x \\ \underline{w}_k^M \end{bmatrix}, \tag{9.15}$$

$$\hat{\underline{y}}_k = \underbrace{\mathbf{H}_k(\underline{\eta}_k^M)\, \underline{x}_k}_{\underline{h}_k(\underline{z}_k)} + \underline{v}_k. \tag{9.16}$$

The nonlinear function $\underline{a}_k^M(\cdot)$ describes the dynamic behavior of the parameters to be estimated. The structure of the augmented model description (9.15) and (9.16) is depicted in Figure 9.12. In this case, it is obvious that the augmented measurement model is *nonlinear* in the augmented state vector $\underline{z}_k$ due to the multiplication of $\mathbf{H}_k(\underline{\eta}_k^M)$ and the system state $\underline{x}_k$; see Example 2. Thus, the node locations $\underline{\eta}_k^M$ characterizes the measurement matrix $\mathbf{H}_k$ and the actual measured values.

It is important to emphasize that the measurement model (9.16) contains a high-dimensional linear substructure, which can be exploited by the application of a more efficient estimator. In the following section, we describe a novel estimator – the *Sliced Gaussian Mixture Filter* – allowing the decomposition of the simultaneous state and parameter estimation. This results in a very efficient localization of the individual nodes in the sensor network.

9.6   Decomposition of the Estimation Problem

In many applications, estimating the state of a system from noisy measurements is a common task. In some special cases, such as *linear systems* with Gaussian random variables, exact solutions to the estimation problem can be found by applying the Kalman filter [82]. There exist a vast variety of approaches for *nonlinear systems* with non-Gaussian random variables. For the efficient implementation of the Bayesian estimator for such systems, the approximation of the true density is necessary. The well-known extended Kalman filter [148] is based on a linearization of the used nonlinear system equations and the application of the standard Kalman filter on the resulting linearized equations. The unscented Kalman filter [78] uses a deterministic sampling of the true density and thus offers an increased higher-order accuracy compared to the extended Kalman filter. However, for both filters the resulting single Gaussian density is often not sufficient for representing the underlying true density. One possibility for increasing the performance is to use a sample representation of the underlying density function, like in particle filters [2, 49].

The mathematical model arising in many applications contains a linear sub–structure. In particular, in the case of the system equations (9.15) and (9.16) for distributed phenomena, the *linear* substructure is *high-dimensional*, and thus offers the chance to significantly increase the estimation performance. There are several methods to solve the combined linear/nonlinear estimation problem while beneficially exploiting the linear substructure. The *decomposition* of the entire estimation problem into a *linear* and a *nonlinear* problem allows for an overall more efficient process. The marginalized particle filter [16, 18, 86, 140] uses a marginalization over the linear sub-space to reduce the dimensionality of the entire state space. The remaining density is subsequently represented by particles. Based on this decomposition, the *standard particle filter* is adopted to cope with the reduced non-linear problem and the *Kalman filter* is exploited in order to find the optimal estimate for the linear sub-space associated with each individual particle. In comparison to the standard particle filter, the marginalized filter certainly improves the performance; however, some drawbacks still remain. For instance, special algorithms are necessary in order to avoid sample degeneration and impoverishment. In addition, measures on how well the true joint density is represented by the estimated one is not provided within this framework.

For that reason, in our previous research work [92, 134], we introduced a more systematic estimator exploiting linear substructures in general nonlinear systems, the so-called *Sliced Gaussian Mixture Filter* (SGMF). There are two key features leading to a significantly improved estimation result.

- **Novel density representation:** The utilization of a special kind of density allows the decomposition of the general estimation problem into a linear and nonlinear problem. To be more specific, as a density representation we employ a so-called *sliced Gaussian mixture density*.

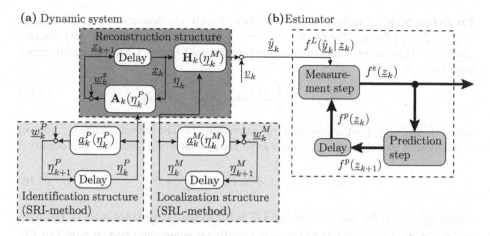

Figure 9.12: Visualization of **(a)** Dynamic System and **(b)** Model-based Bayesian Estimator for the Node Localization based on Local Observations. The system description contains a high-dimensional linear substructure. The node location $\underline{\eta}_k^M$ characterizes the measurement matrix $\mathbf{H}_k$, and thus, the individual measurements(adapted from [132]).

- **Systematic approximation:** The systematic approximation of the density resulting from the estimation update leads to (close to) optimal results. Thus, less parameters for the density description are necessary and a measure for the approximation performance is provided.

Despite the high-dimensional nonlinear character, the systematic approach of the simultaneous state and parameter estimation for large-area distributed phenomena is feasible thanks to a decomposition based on the sliced Gaussian mixture density. Furthermore, the uncertainties occurring in the mathematical system description and arising from noisy measurements are considered by an integrated treatment.

9.6.1 General Prediction and Measurement Step

Before the framework of the Sliced Gaussian Mixture Filter is described in more detail, in this section, we explain the *general prediction and measurement step* for general systems characterized by *nonlinear equations*. The aim of the Bayesian estimator is to calculate the probability density function $f(\underline{z}_k)$ representing the state vector $\underline{z}_k$ as precisely as possible at every time step. Due to its high computational demand and the resulting non-parametric density representation, approximation approaches for the nonlinear estimation problem are inevitable.

In general, the model-based Bayesian estimator consists of two steps being performed alternately in order to derive an estimate for the state vector $\underline{z}_k$ in terms of the density function $f(\underline{z}_k)$. The two steps of the Bayesian estimator are the *prediction step* and the *measurement step*.

Prediction Step   The purpose of the prediction step is to determine, for a given prior density $f^e(\underline{z}_k)$ for $\underline{z}_k$, the predicted density $f^p(\underline{z}_{k+1})$ of $\underline{z}_{k+1}$ for the next discrete time step. This can be achieved by evaluating the well-known Chapman-Kolmogorov equation

$$f^p(\underline{z}_{k+1}) = \int_\Omega f^T(\underline{z}_{k+1}|\underline{z}_k) f^e(\underline{z}_k) \mathrm{d}\underline{z}_k, \qquad (9.17)$$

where $f^T(\underline{z}_{k+1}|\underline{z}_k)$ is the so-called *transition density*. The additive noise $\underline{w}_k^z$ subject to the system model (9.15) is assumed to be zero-mean white Gaussian with density

$$\underline{w}_k^z \sim f_k^w(\underline{w}_k^z) = \mathcal{N}\left(\underline{w}_k^z - \underline{\mu}_k^w, \mathbf{C}_k^w\right) \qquad (9.18)$$

where $\underline{\mu}_k^w = 0$ is the mean vector and $\mathbf{C}_k^w$ is the system covariance matrix. The transition density $f^T(\cdot)$ strongly relies on the nonlinear augmented system model (9.15). Assuming the system noise $\underline{w}_k^z$ to be given by (9.18), the transition density $f^T(\cdot)$ can be derived according to

$$f^T(\underline{z}_{k+1}|\underline{z}_k) = \mathcal{N}\left(\underline{z}_{k+1} - \underline{a}_k(\underline{z}_k, \hat{\underline{u}}_k), \mathbf{C}_k^w\right),$$

which characterizes the probability of the transition of the state vector $\underline{z}_k$ to the next time step. It is clear that the structure of the transition density $f^T(\cdot)$ strongly depends on the actual structure of the underlying system model.

Measurement Step   The information of measurements $\hat{\underline{y}}_k$ can be incorporated into the processing scheme in order to improve the estimation of $\underline{z}_k$. The estimated density $f^e(\underline{z}_k)$ can be determined by the famous Bayes' formula

$$f^e(\underline{z}_k) = c \cdot f^L(\hat{\underline{y}}_k|\underline{z}_k) \cdot f^p(\underline{z}_k), \qquad (9.19)$$

where the coefficient c is a normalization constant. The density function $f^L(\underline{y}_k|\underline{z}_k)$ is the so-called *likelihood*. The additive noise $\underline{v}_k$ subject to the measurement model (9.16) is assumed to be zero-mean white Gaussian with density

$$\underline{v}_k \sim f_k^w(\underline{v}_k) = \mathcal{N}\left(\underline{v}_k - \underline{\mu}_k^v, \mathbf{C}_k^v\right), \qquad (9.20)$$

where $\underline{\mu}_k^v = 0$ is the mean vector and $\mathbf{C}_k^v$ is the measurement covariance matrix. Assuming the measurement noise $\underline{v}_k$ to be given by (9.20), the likelihood $f^L(\cdot)$ can be derived according to

$$f^L(\hat{\underline{y}}_k|\underline{z}_k) = \mathcal{N}\left(\hat{\underline{y}}_k - \underline{h}_k(\underline{z}_k), \mathbf{C}_k^v\right),$$

which can be regarded as the conditional density for the occurrence of the measurement $\hat{\underline{y}}_k$ for given $\underline{x}_k$. It is obvious that the structure of the likelihood function $f^L(\cdot)$ depends on the structure of the measurement model.

9.6.2   The Sliced Gaussian Mixture Filter (SGMF)

In this section, we describe the Bayesian prediction step and measurement step based on *sliced Gaussian mixture densities*. It is shown, how this novel density representation can be exploited for decomposing the general prediction step (9.17) and measurement step (9.19) into a *linear* and *nonlinear part*. By this means, a more efficient closed-form calculation of the simultaneous state and parameter estimation of distributed phenomena is possible. The framework of decomposing the nonlinear estimation problem can be applied to various dynamic systems. However, we restrict our attention to the augmented system model (9.15) and measurement model (9.16) necessary for the node localization (SRL method) [132]. The resulting equations of the prediction step and measurement step in the case of the identification of model parameters (SRI method) can be found in [134].

The *Sliced Gaussian Mixture Filter* basically consists of three steps: the *decomposition* of the estimation problem, the utilization of an *efficient update*, and the *re-approximation* of the density representation.

1. **Decomposition:** The nonlinear high-dimensional estimation problem is decomposed into a *linear* high-dimensional problem (state estimation) and *nonlinear* low-dimensional problem (parameter estimation). This can be achieved by means of the *sliced Gaussian mixture density*; see Figure 9.13 (**a**).

2. **Efficient update:** Based on the decomposition in a linear and nonlinear estimation problem both the prediction step and the measurement step can be performed with an overall more efficient performance. Basically, analytic and efficient estimators, such as Kalman filter, are exploited to efficiently perform the estimation update in the linear high-dimensional subspace. The estimation in the nonlinear subspace is performed by nonlinear estimators, such as the Dirac mixture filter, visualized in Figure 9.13 (**b**).

3. **Re-approximation:** The estimation based on the sliced Gaussian mixture densities leads to a density representation consisting of *Gaussian mixtures* in all subspaces. In order to bound the complexity, the resulting density needs to be re-approximated by means of the *sliced Gaussian mixture density*, depicted in Figure 9.13 (**b**).

The remainder of this section is devoted to a more detailed description on the three steps of the Sliced Gaussian Mixture Filter (SGMF). Furthermore, the update equations for the simultaneous reconstruction and node localization (SRL method) are given.

Decomposition through sliced Gaussian mixtures   For the decomposition of the nonlinear estimation problem into a (conditionally) linear and a nonlinear problem, we proposed in our previous research work [92] [134] so-called *sliced Gaussian mixture densities* as a density representation. This density function $f(\underline{x}_k, \underline{\eta}_k)$ is represented by a Dirac mixture in the *nonlinear* sub-space $\underline{\eta}_k$ (parameter space) and Gaussian mixture

(a) Sliced Gaussian mixture densities   **(b)** Efficient update and reapproximation

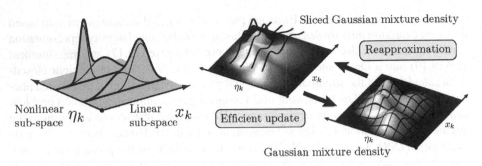

Figure 9.13: **(a)** Visualization of a *Sliced Gaussian Mixture Density* Consisting of a Gaussian Mixture in x_k Subspace and a Dirac Mixture in η_k Subspace. **(b)** The efficient update and the re-approximation of the resulting *Gaussian mixture* density by a *sliced Gaussian mixture* density is alternately performed(adapted from [132]).

in the linear sub-space $\underline{x}_k$ (state space),

$$f(\underline{x}_k, \underline{\eta}_k) = \sum_{i=1}^{M} \alpha_k^i \underbrace{\delta(\underline{\eta}_k - \underline{\xi}_k^i)}_{\text{Dirac mixture}} \cdot \underbrace{f(\underline{x}_k | \underline{\xi}_k^i)}_{\text{Gaussian mixture}}, \tag{9.21}$$

where $\delta(\cdot)$ is the Dirac delta function and α_k^i their respective weighting coefficients. The density parameters $\underline{\xi}_k^i \in \mathbb{R}^s$ can be regarded as the position of the individual density slices, as shown in Figure 9.13 **(a)**. The marginal density in nonlinear sub-space $\underline{\eta}_k$ is given by a Dirac mixture function, according to

$$f(\underline{\eta}_k) = \sum_{i=1}^{M} \alpha_k^i \delta(\underline{\eta}_k - \underline{\xi}_k^i), \tag{9.22}$$

where α_k^i and $\underline{\xi}_k^i$ represent the weights and positions of the Dirac functions, respectively. The density representation along the individual slices is assumed to be a Gaussian mixture density

$$f(\underline{x}_k | \underline{\xi}_k^i) = \sum_{j=1}^{N^i} \beta_k^{ij} \mathcal{N}\left(\underline{x}_k - \underline{\mu}_k^{ij}, \mathbf{C}_k^{ij}\right), \tag{9.23}$$

with $\beta_k^{ij}, \underline{\mu}_k^{ij} \in \mathbb{R}^r$, and $\mathbf{C}_k^{ij} \in \mathbb{R}^{r \times r}$ denoting the weights, means, and covariance matrices of the j-th component of the Gaussian mixture density of the i-th slice.

Efficient update and posterior density   Thanks to the system model (9.15) and *conditionally linear* measurement model (9.16), the Chapman-Kolmogorov equation for the prediction step and the Bayes formula for the measurement step can be solved analytically. The proof is omitted here, only the resulting predicted density is stated.

By means of the *sliced Gaussian mixture filter*, the predicted density $\widetilde{f}^p$ results in a Gaussian mixture in linear $\underline{x}_k$ and nonlinear sub-space $\underline{\eta}_k^M$,

$$\widetilde{f}^p(\underline{x}_{k+1}, \underline{\eta}_{k+1}^S) = c \cdot \sum_{i=1}^{M} \sum_{j=1}^{N^i} \alpha_k^i \beta_k^{ij} \gamma_k^{ij}$$
$$\cdot \, \mathcal{N}\left(\underline{\eta}_{k+1}^S - \underline{\xi}_{k+1}^{pi}, \mathbf{C}_w^n\right) \mathcal{N}\left(\underline{x}_{k+1} - \underline{\mu}_{k+1}^{pij}, \mathbf{C}_{k+1}^{pij}\right), \tag{9.24}$$

where the mean $\underline{\mu}_{k+1}^{pij}$ and covariance matrices $\mathbf{C}_{k+1}^{pij}$ in linear sub-space $\underline{x}_k$ are calculated by applying the standard Kalman filter. The mean in nonlinear sub-space $\underline{\eta}_k^M$ is derived by simply repositioning the density slices.

In the following, the parameters of the posterior density (9.24) for the node localization in a sensor network (SRL method) are stated. For the sake of simplicity and in order to keep the equations simple, the abbreviation $\mathbf{H}_k^i := \mathbf{H}(\underline{\xi}_k^i)$ is used.

- Parameters of resulting density (9.24) for the *prediction step*:

$$\text{Mean vectors} \qquad \underline{\mu}_{k+1}^{pij} := \mathbf{A}_k \underline{\mu}_k^{eij} + \mathbf{B}_k \underline{\hat{u}}_k$$

$$\text{Covariance matrices} \qquad \mathbf{C}_{k+1}^{pij} := \mathbf{A}_k \mathbf{C}_k^{eij} \mathbf{A}_k^T + \mathbf{C}_w^l$$

$$\text{Positions in nonlinear subspace} \qquad \underline{\xi}_{k+1}^{pi} := \underline{a}_k^M\left(\underline{\xi}_k^{ei}\right)$$

- Parameters of resulting density (9.24) for the *measurement step*:

$$\text{Weights of the slice} \qquad \gamma_k^{ij} := \mathcal{N}\left(\underline{\hat{y}}_k - \mathbf{H}_k^i \underline{\mu}_k^{pij}, \mathbf{H}_k^i \mathbf{C}_k^{pij} \mathbf{H}_k^{i\,T} + \mathbf{C}_k^v\right)$$

$$\text{Mean vectors} \qquad \underline{\mu}_k^{eij} := \underline{\mu}_k^{pij} + \mathbf{K}\left(\underline{\hat{y}}_k - \mathbf{H}_k^i \underline{\mu}_k^{pij}\right)$$

$$\text{Covariance matrices} \qquad \mathbf{C}_k^{eij} := \mathbf{C}_k^{pij} - \mathbf{K}\mathbf{H}_k^i \mathbf{C}_k^{pij}$$

$$\text{Kalman gains} \qquad \mathbf{K} := \mathbf{C}_k^{pij} \mathbf{H}_k^{i\,T}\left(\mathbf{C}_k^v + \mathbf{H}_k^i \mathbf{C}_k^{pij} \mathbf{H}_k^{i\,T}\right)^{-1}$$

At this point it is important to emphasize that the aforementioned equations for the parameters of the posterior density (9.24) strongly depend on the actual model structure (9.15) and (9.16). For the simultaneous reconstruction and identification of distributed phenomena (SRI method) similar equations can be found, as it was derived in our previous research work [134].

Re-approximation and bounding complexity In order to bound the complexity, the predicted density (9.24) in terms of a Gaussian mixture density needs to be re-approximated by a sliced Gaussian mixture density (9.21). There are several approaches to perform this approximation. One possible approach for the approximation is to derive the location of the density slices by only considering the marginal density $\widetilde{f}^p(\underline{\eta}_{k+1}^M)$. In general, the approximation of arbitrary marginal densities by Dirac mixture densities (9.22) can be achieved by: *batch approximation* [141] or *sequential approximation* [52].

The *batch approximation* is an efficient solution procedure for arbitrary true density functions on the basis of homotopy continuation (progressive Bayes). This procedure results in an optimal solution. The *sequential approximation* is based on inserting one component of the Dirac mixture density at a time. The key idea of this algorithm is that every component of the Dirac mixture density corresponds to an interval in the nonlinear subspace of the sample space and approximates the true marginal density only in the corresponding interval. Then, based on the splitting of the intervals and their respective component of the Dirac mixture density arbitrary densities can be approximated.

After the approximation of the marginal density $\widetilde{f}^p(\underline{\eta}_{k+1}^M)$ in the nonlinear subspace, the Dirac approximation is extended to a sliced Gaussian mixture representation over the entire sample space. Basically, this is achieved by evaluating the Gaussian mixture density $\widetilde{f}^p(\underline{x}_{k+1}, \underline{\eta}_{k+1}^M)$ at every Dirac position. This leads to a sliced Gaussian mixture density (9.21), which can be used for the next processing step. A more detailed description on the re-approximation can be found in [92].

9.7   Application: Node Localization

In this section, we demonstrate the application and performance of the proposed sensor node localization method (SRL method). As it is described above, the localization problem is restated as a *simultaneous state and parameter estimation* problem. The resulting high-dimensional nonlinear problem is decomposed into a linear and nonlinear part by means of the *Sliced Gaussian Mixture Filter*, and thus leads to an overall more efficient localization method.

There are four key features characterizing the novelties of the proposed method for the passive localization (SRL method):

- The approach is based on *local measurements* of distributed phenomena *only*.

- The *uncertainties* in the mathematical model and the local measurements of the sensor network are *systematically* considered.

- For the estimated node locations, an *uncertainty measure* is derived in terms of a density function.

- The *simultaneous approach* allows improving the estimation of the distributed phenomenon, which then can be exploited for localizing other nodes.

For a more detailed description of the simultaneous reconstruction and node localization method (SRL method), the interested reader is referred to [132].

By means of simulation results, we investigate the accuracy of the identified location $\underline{\eta}_k^M$ of a sensor node *locally* measuring a distributed phenomenon. The underlying distributed phenomenon is assumed to be given as follows:

Example 6 (Simulated system)

In this simulation, we consider the localization of sensor nodes based on the one-dimensional partial differential equation (9.5). The initial conditions and Dirichlet boundary conditions are depicted in Figure 9.7 **(c)**–**(d)**. Here, we assume that the sensor

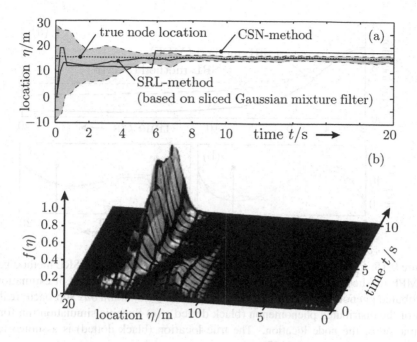

Figure 9.14: Comparison of SRL method based on SGMF, SRL method based on MPF, and Deterministic Approach CSN-method. **(a)** Root mean square error over time of 100 simulation runs. **(b)** Specific density function $f(\eta_k^M)$ for estimated node location η_k^M over time(adapted from [132]).

network consists only of one single sensor node locally measuring the phenomenon. Furthermore, the sensor node has only very uncertain knowledge about the initial distributed phenomenon; see Figure 9.7 **(c)**. The goal is to localize the sensor node with initially unknown location using local measurements of the distributed phenomenon. The true node location is assumed to be $\eta_{\text{true}}^S = 16\,\text{m}$. The system noise term is $C_l^w = \text{diag}\{20,\ldots,20\}$, the noise term for the node location is given by $C_n^w = 0.03$, and for the local measurement of the node to be localized is assumed to be $C_k^v = 0.01$. Here, we compare different approaches for the passive localization:

- Deterministic approach introduced in [66] (CSN-method)
- SRL method based on the *sliced Gaussian mixture filter* (using 50 slices)
- SRL method based on the *marginalized particle filter* (using 500 particles)

The aforementioned approaches for node localizations are compared based on 100 Monte-Carlo simulation runs. In particular, the accuracy of the estimated location η_k^M is investigated. The results of the localization methods are shown in Figure 9.14 and Figure 9.15.

The estimation of the unknown location η_k^M for one specific simulation run is depicted in Figure 9.14 **(a)**. It can be clearly seen that after a certain transition time the SRL method based on sliced Gaussian mixture filter (with 50 slices) offers a nearly exact location. The estimation of the deterministic approach (CSN-method) strongly

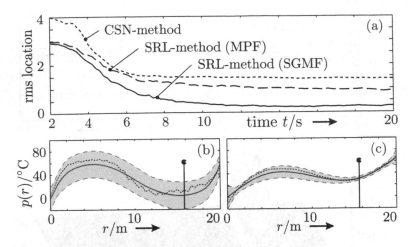

Figure 9.15: Comparison of SRL method based on SGMF (black), SRL method based on MPF (dashed), and CSN-method (dotted). **(a)–(b)** Improvement of estimation of distributed phenomenon (gray shaded area) thanks to *simultaneous* approach; realization of the distributed phenomenon (black dotted). **(c)** Specific simulation run for the estimation of the node location. The true location (black dotted) is assumed to be $\eta^{M}_{\text{true}} = 16\,\text{m}$(adapted from [132]).

deviates from the true location. This is caused mainly by neglected system and measurement noises. The entire density function $f(\eta^{M}_{k})$ for the estimated location η^{M}_{k} for a specific simulation run is depicted in Figure 9.14 **(b)**. The ambagious distribution of the physical phenomenon to be observed results in a multimodal density function for the estimated location $\underline{\eta}^{M}_{k}$. This explains the higher uncertainty at the beginning of the simulation. However, by exploiting more and more measurements and information about the dynamic system, the estimation of the location changes from a multimodal to an unimodal function. Thus, the location estimate becomes more accurate and certain.

The root mean square error (rms) of all 100 simulation runs over time is depicted in Figure 9.15 **(a)**. It is obvious that the SRL method based on the Sliced Gaussian Mixture Filter (with 50 slices) outperforms both the deterministic approach (CSN-method) and the approach based on marginalized particle filter (with 500 particles). This is mainly due to the stochastic approach and the systematic approximation of the density used in the framework of the simultaneous state and parameter estimation.

Thanks to the *simultaneous property* of the SRL method, not only can the sensor node be accurately localized, but also the estimate of the distributed phenomenon can be further improved. The improvement of the reconstruction result is obvious by comparing Figure 9.14 **(b)** with Figure 9.14 **(c)**. It is important to emphasize that the phenomenon can be reconstructed at the actual measurement point as well as at non-measurement points. The improved knowledge about the phenomenon in the entire solution domain can be exploited by other sensor nodes to localize themselves.

In this work, we restricted ourselves to the localization of one single sensor node locally measuring a distributed phenomenon. It is believed that using more than one sensor node, the performance of the localization process can be significantly improved since more information about the distributed phenomenon can be exploited. Furthermore, already localized sensor nodes, e.g., sensor beacons or base stations, can be used to reconstruct the physical phenomenon, and thus support the localization of individual sensor nodes deployed between the beacons.

9.8 Conclusions and Future Work

In this chapter, we describe the methodology for the *simultaneous state and parameter estimation* of *distributed phenomena*. The spatial and temporal decomposition of the distributed system results in a finite-dimensional model in state space form (usually characterized by a high-dimensional state vector). Hence, the augmentation of the system state with the parameter to be estimated leads to a high-dimensional nonlinear system description. Based on a novel density representation – *sliced Gaussian mixture density* – the linear sub-structure contained in the finite-dimensional model is exploited. This leads to an overall more efficient estimation process. The performance is demonstrated by means of simulation results and it turns out that, compared to other nonlinear estimators, the *sliced Gaussian mixture filter* achieves a higher accuracy.

The application of the proposed method for the simultaneous state and parameter estimation to sensor networks provides novel prospects. The network is able to estimate the *entire* state of the distributed phenomenon, identify *non-measurable* quantities, verify and validate the correctness of the estimation results, and adapt autonomously their algorithms. Within the proposed framework, a novel method for the localization of individual sensor nodes is introduced. The localization method (SRL method) performs without relying on a satellite positioning system (which is not always available, e.g., indoor applications) as long as a strong model of the surrounding is available.

So far, the model parameters and structure were assumed to be precisely known for the SRL method. In many real world applications, however, the parameters contain remaining uncertainties, or even could be completely unknown. The combination of the parameter identification of distributed phenomena and the node localization is left for future research work. Finally, it is intended to test the proposed localization methods on actual sensor data.

For the observation of large-area distributed phenomena, decentralized methods are inevitable in order to cope with high-dimensional state vectors. Hence, further decompositions both in the linear subspace and nonlinear subspace are necessary, similar to [131]. This is left for future research work.

Bibliography

[1] A. Adamatzky, B.D.L. Cotello, and T. Asai. *Reaction-Diffusion Computers*. Elsevier, Amsterdam, The Netherlands, 2005.

[2] M. S. Arulampalam, S. Maskell, N. Gordon, and T. Clapp. A Tutorial on Particle Filters for Online Nonlinear/Non-Gaussian Bayesian Tracking. *IEEE Transactions on Signal Processing*, 50(2):174–188, February 2002.

[3] E. A. Ashcroft. R for Semantics. *ACM Transactions on Programming Languages and Systems*, 4(2):283–295, 1982.

[4] A. Babloyantz and J. Hiernaux. Models for Cell Differentiation and Generation of Polarity in Diffusion-Governed Morphogenetic Fields. *Bulletin of Mathematical Biology*, 37:637–657, 1975.

[5] T. Bader, A. Wiedemann, K. Roberts, and U.D. Hanebeck. Model-based Motion Estimation of Elastic Surfaces for Minimally Invasive Cardiac Surgery. In *IEEE International Conference on Robotics and Automation (ICRA 2007)*, Rome, Italy, April 2007.

[6] A. J. Baker. *Finite Element Computational Fluid Mechanics*. Taylor and Francis, London, UK, 1983.

[7] J. Bard and I. Lauder. How Well does Turing's Theory of Morphogenesis Work? *Jnl Theor Biology*, 45:501–531, 1974.

[8] J.E. Bares and D.S. Wettergreen. Dante II: Technical Description, Results, and Lessons Learned. *The International Journal of Robotics Research*, 18(7):621–649, July 1999.

[9] P. Bartelt and M. Lehning. A Physical SNOWPACK Model for the Swiss Avalanche Warning, Part I: Numerical Model. *Cold Regions Science and Technology*, 35:123–145, 2002.

[10] B. Bhanu. Evaluation of Automatic Target Recognition Algorithms. In P.-E. Danielsson and A.J. Oosterlinck, editors, *Proceedings Vol. 435 SPIE Conference on Architecture and Algorithms for Digital Image Processing*, pages 18–25, August 1983.

[11] P. Bonnet, J.E. Gehrke, and P. Seshadri. Towards Sensor Database Systems. In *Proc of the Second Intntl Conf on Mobile Data Management*, Hong Kong, January 2001.

[12] A.J. Briggs and B.R. Donald. Automatic Sensor Configuration for Task-directed Planning. In *IEEE Conference on Robotics and Automation*, pages 1345–1350, San Diego, CA, May 1994.

[13] R.R. Brooks and S. Iyengar. Averaging Algorithm for Multi-dimensional Redundant Sensor Arrays: Resolving Sensor Inconsistencies. Technical report, Louisiana State University, Baton Rouge, LA, 1993.

[14] D. Brunn, F. Sawo, and U.D. Hanebeck. Modellbasierte Vermessung verteilter Phänomene und Generierung optimaler Messsequenzen. *tm - Technisches Messen, Oldenbourg Verlag*, 3:75–90, March 2007.

[15] N. Bulusu, D. Estrin, L. Girod, and J. Heidemann. Scalable Coordination for Wireless Sensor Networks: Self-Configuring Localization Systems. In *Proc. Sixth International Symposium on Communication Theory and Applications (ISCTA '01)*, Ambleside, Lake District, UK, July 2001.

[16] G. Casella and C.P. Robert. Rao-Blackwellisation of Sampling Schemes. *Biometrika*, 83(1):81–94, 1996.

[17] H. Chan and A. Perrig. ACE: An Emergent Algorithm for Highly Uniform Cluster Formation. In *Proceedings of First European Workshop on Wireless Sensor Networks*, Berlin, Germany, January 2004.

[18] R. Chen and J.S. Liu. Mixture Kalman Filters. *Journal of the Royal Statistical Society*, 62(3):493–508, 2000.

[19] Y. Chen. SNETs: Smart Sensor Networks. Master's thesis, University of Utah, Salt Lake City, Utah, December 2000.

[20] Y. Chen and T.C. Henderson. S-Nets: Smart Sensor Networks. In *Proceedings of the International Symposium on Experimental Robotics*, pages 85–94, Honolulu, Hawaii, December 2000.

[21] C.-Y. Chong and S.P. Kumar. Sensor Networks: Evolution, Opportunities, and Challenges. *Proceedings of the IEEE*, 91(8):1247–1256, 2003.

[22] T.J. Chung. *Computational Fluid Dynamics*. Cambridge University Press, Cambridge, UK, 2002.

[23] P. Corke, R. Peterson, and D. Rus. Localization and Navigation Assisted by Cooperating Networked Sensors and Robots. *International Journal of Robotics Research*, 24(9):771–786, September 2005.

[24] D. Culler, P. Dutta, C.T. Ee, R. Fonseca, J. Hui, P. Levis, J. Polastre, S. Shenker, I. Stoica, G. Tolle, and J. Zhao. Towards a Sensor Network Architecture: Lowering the Waistline. In *HOTOS'05: Proceedings of the 10th conference on Hot Topics in Operating Systems*, pages 24–24, Berkeley, CA, USA, 2005.

[25] D. Culler, D. Estrin, and M. Srivastava. Overview of Sensor Networks. *IEEE Computer*, 37(8):41–49, 2004.

[26] M. Dekhil. *Instrumented Logical Sensor Systems*. PhD thesis, University of Utah, Salt Lake City, UT, August 1998.

[27] M. Dekhil and T.C. Henderson. Instrumented Sensor Systems. In *IEEE International Conference on Multisensor Fusion and Integration (MFI 96)*, pages 193–200, Washington, D.C., December 1996.

[28] M. Dekhil and T.C. Henderson. Optimal Wall Pose Determination in a Shared-Memory Multi-Tasking Control Architecture. In *IEEE International Conference on Multisensor Fusion and Integration (MFI 96)*, pages 736–741, Washington, D.C, December 1996.

[29] M. Dekhil and T.C. Henderson. Instrumented Sensor System Architecture. UUCS-97-011, University of Utah, Salt Lake City, UT, March 1997.

[30] M. Dekhil and T.C. Henderson. Instrumented Sensor System Architecture. Technical Report UUCS-97-011, University of Utah, Department of Computer Science, August 1997.

[31] M. Dekhil and T.C. Henderson. Instrumented Logical Sensors Systems. *International Journal of Robotics Research*, 17(4):402–417, 1998.

[32] M. Dekhil and T.M. Sobh. Embedded Tolerance Analysis for Sonar Sensors. In *Invited paper to the special session of the 1997 Measurement Science Conference, Measuring Sensed Data for Robotics and Automation*, Pasadena, CA, January 1997.

[33] J.L. Devore. *Probability and Statistics for Engineering and the Sciences*. Duxbury Press, Pacific Grove, CA, 1995.

[34] B.R. Donald. On Information Invariants in Robotics. *Artificial Intelligence*, 72:217–304, 1995.

[35] H.F. Durrant-Whyte. *Integration, Coordination and Control of Multisensor Robot Systems*. Kluwer Academic Publishers, Boston, MA, 1988.

[36] D. Estrin, R. Govindan, J. Heidemann, and S. Kumar. Next Century Challenges: Scalable Coordination in Sensor Networks. In *Proceedings of Mobicom 1999*, Seattle, WA, August 1999.

[37] H.R. Everett. *Sensors for Mobile Robots Therory and Application*. A K Peters, Ltd., Massachusetts, 1995.

[38] W.S. Fai. A Multi-sensor Integration and Data Acquisition System. Master's thesis, University of Utah, Salt Lake City, Utah, June 1983.

[39] O. Faugeras. *Three-dimensional Computer Vision - A Geometric viewpoint*. The MIT Press, Cambridge, MA, 1993.

[40] D. Floreano, J.D. Nicoud, and F. Mondada, editors. *Advances in Artificial Life, 5th European Conference, ECAL'99, Lausanne, Switzerland, September 13-17, 1999, Proceedings*, volume 1674 of *Lecture Notes in Computer Science*. Springer, 1999.

[41] J. Fraden. *AIP Handbook of Modern Sensors*. American Institute of Physics, New York, 1993.

[42] E. Gamma, R. Helm, R. Johnson, and J. Vlissides. *Design Patterns: Elements of Reusable Object-oriented Software*. Addison Wesley, Reading, MA, 1995.

[43] D. Ganesan, D. Estrin, A. Woo, and D. Culler. Complex Behavior at Scale: An Experimental Study of Low-Power Wireless Sensor Networks. Technical Report CSD-TR 02-0013, University of California at Los Angeles, Department of Computer Science, February 2002.

[44] C. Giraud and B. Jouvencel. Sensor Selection in a Fusion Process: a Fuzzy Approach. In R.C. Luo, editor, *Proceedings of the IEEE International Conference on Multisensor Fusion and Integration*, pages 599–606, Las Vegas, NV, 1994.

[45] M.J.C. Gordon. *Denotational Description of Programming Languages*. Springer-Verlag, New York, NY, 1979.

[46] M.I. Granero, A. Porati, and D. Zanacca. A Bifurcation Analysis of Pattern Formation in a Diffusion Goverened Morphogenetic Field. *Jnl of Mathematical Biology*, 4:21–27, 1977.

[47] M. Grigoras, O. Feiermann, and U.D. Hanebeck. Data-Driven Modeling of Signal Strength Distributions for Localization in Cellular Radio Networks (Datengetriebene Modellierung von Feldstärkeverteilungen für die Ortung in zellulären Funknetzen). *at - Automatisierungstechnik - Automatisierungstechnik, Sonderheft: Datenfusion in der Automatisierungstechnik*, 53(7):314–321, July 2005.

[48] F.C.A. Groen, P.P.J. Antonissen, and G.A. Weller. Model Based Robot Vision. In *IEEE Instrumentation and Measurement Technology Conference*, pages 584–588, Irvine, CA, May 1993.

[49] F. Gustafsson, F. Gunnarsson, N. Bergman, U. Forssell, J. Jansson, R. Karlsson, and P.-J. Nordlund. Particle Filters for Positioning, Navigation, and Tracking. *IEEE Transactions on Signal Processing*, 50(2):425–437, 2002.

[50] G.D. Hager and M. Mintz. Task-directed Multisensor Fusion. In *IEEE Conference on Robotics and Automation*, pages 662–667, Scottsdale, AZ, 1989.

[51] G.D. Hager and M. Mintz. Computational Methods for Task-directed Sensor Data Fusion and Sensor Planning. *Int. J. Robotics Research*, 10(4):285–313, August 1991.

[52] U.D. Hanebeck and O.C. Schrempf. Greedy Algorithms for Dirac Mixture Approximation of Arbitrary Probability Density Functions. In *IEEE Conference on Decision and Control (CDC 2007)*, New Orleans, LA, December 2007.

[53] W. Heizelman, A. Chandrakasan, and H. Balakrishnan. Energy Efficient Communication Protocol for Wireless Microsensor Networks. In *Proceedings of 33rd Hawaii International Conference on System Sciences*, Hawaii, January 2000.

[54] T. Henderson, W. Fai, and C. Hansen. MKS: A Multisensor Kernel System. *IEEE Trans. Systems Man and Cybernetics*, 14(5):784–791, 1984.

[55] T.C. Henderson. Leadership Protocol for S-Nets. In *Proc Multisensor Fusion and Integration*, pages 289–292, Baden-Baden, Germany, August 2001.

[56] T.C. Henderson. Verification and Validation of Sensor Networks. In *Schloss Dagstuhl Workshop on Form and Content of Sensor Networks*, Wadern, Germany, September 2005.

[57] T.C. Henderson. Performance Measures for Sensor Networks. In *NIST-ANS Workshop on Urban Search and Rescue Performance Measures for Intelligent Systems*, Salt Lake City, UT, February 2006.

[58] T.C. Henderson. Further Observations on the SNL Wireless Sensor Network Leadership Protocol. In Sukhan Lee, editor, *Proceedings of the IEEE International Conference on Multisensor Fusion and Integration*, Seoul, South Korea, 2008.

[59] T.C. Henderson, B. Bruderlin, M. Dekhil, L. Schenkat, and L. Veigel. Sonar Sensing Strategies. In *IEEE Conference on Robotics and Automation*, pages 341–346, Minneapolis, MN, April 1996.

[60] T.C. Henderson and M. Dekhil. Visual Target Based Wall Pose Estimation. Technical Report UUCS-97-010, University of Utah, Department of Computer Science, Salt Lake City, UT, July 1997.

[61] T.C. Henderson, M. Dekhil, B. Bruderlin, L. Schenkat, and L. Veigel. Flat Surface Recovery from Sonar Data. In *DARPA Image Understanding Workshop*, pages 995–1000, Palm Springs, CA, February 1996.

[62] T.C. Henderson, M. Dekhil, S. Morris, Y. Chen, and W.B. Thompson. Smart Sensor Snow. *IEEE Conference on Intelligent Robots and Intelligent Systems*, October 1998.

[63] T.C. Henderson and E. Grant. Gradient Calculation in Sensor Networks. In *Proc International Conf on Intelligent Robots and Systems*, Sendai, Japan, October 2004.

[64] T.C. Henderson, J.-C. Park, N. Smith, and R. Wright. From Motes to Java Stamps: Smart Sensor Network Testbeds. In *Proc International Conf on Intelligent Robots and Systems*, Las Vegas, NV, October 2003.

[65] T.C. Henderson and E. Shilcrat. Logical Sensor Systems. *Journal of Robotic Systems*, 1(2):169–193, 1984.

[66] T.C. Henderson, C. Sikorski, K. Luthy, and E. Grant. Computational Sensor Networks. In *Proc International Conf on Intelligent Robots and Systems*, San Diego, CA, October 2007.

[67] T.C. Henderson, R. Venkataraman, and G. Choikim. Reaction-Diffusion Patterns in Smart Sensor Networks. In *Proc International Conference on Robotics and Automation*, New Orleans, April 2004.

[68] J. Hightower and G. Borriello. Location Systems for Ubiquitous Computing. *IEEE Computer*, 34(8), 2001.

[69] J. Hill and D. Culler. A Wireless Embedded Sensor Architecture for System-Level Optimization. ECE, UC Berkeley, October 2002.

[70] J.P. Holman and Jr. W.J. Gajda. *Experimental Methods for Engineers*. McGraw-Hill, New York, NY, 1978.

[71] H. Hu, J.M. Brady, F. Du, and P.J. Probert. Distributed Real-time Control of a Mobile Robot. *Jnl of Intelligent Automation and Software Computing*, pages 63–83, August 1995.

[72] T. Imielinski and S. Goel. DataSpace - Querying and Monitoring Deeply Networked Collections in Physical Space. In *Proc. of International Workshop on Data Engineering for Wireless and Mobile Access (MobiDE'99)*, Seattle, WA, August 1999.

[73] C. Intanagonwiwat, R. Govindan, and D. Estrin. Directed Diffusion: A Scalable and Robust Communication Paradigm for Sensor Networks. In *Proc. of Mobicom 2000*, Boston, August 2000.

[74] S.S. Iyengar and L. Prasad. A General Computational Framework for Distributed Sensing and Fault-tolerant Sensor Integration. *IEEE Transactions on Systems, Man and Cybernetics*, 25(4):643–650, April 1995.

[75] A. Jeremic and A. Nehorai. Design of Chemical Sensor Arrays for Monitoring Disposal Sites on the Ocean Floor. *IEEE Journal of Oceanic Engineering*, 23:334–343, 1998.

[76] J. Jin, X. Gao, S. Sorooshian, Z.-L. Yang, R. Bales, R.E. Dickinson, S.-F. Sun, and G.-X. Wu. One-dimensional Snow Water and Energy Balance Model for Vegetated Surfaces. *Hydrological Processes*, 13:2467 – 2482, 1999.

[77] R. Joshi and A.C. Sanderson. Model-based Multisensor Data Fusion: a Minimal Representation Approach. In *IEEE Conference on Robotics and Automation*, pages 477–484, San Diego, CA, May 1994.

[78] S.J. Julier and J.K. Uhlmann. Unscented Filtering and Nonlinear Estimation. *Proceedings of the IEEE*, 92(3):401–422, 2004.

[79] E.W. Justh and P.S. Krishnaprasad. Pattern-forming Systems for Control of Large Arrays of Actuators. *Jnl of Nonlinear Sci*, 11(4):239–277, 2001.

[80] G. Kahn. The Semantics of a Simple Language for Parallel Programming. In *Proceedings of IFIP*, 1974.

[81] G. Kahn and D. MacQueen. Coroutines and Networks of Parallel Processes. In *Proceedings of IFIP*, 1974.

[82] R.E. Kalman. A New Approach to Linear Filtering and Prediction Problems. *Transactions of the ASME - Journal of Basic Engineering*, 82:35–45, 1960.

[83] T. Kang, C.R. Merritt, B. Karaguzel, J.M. Wilson, P.D. Franzon, B. Pourdeyhimi, E. Grant, and T. Nagle. Sensors on Textile Substrates for Home-Based Healthcare Monitoring. In *Conference on Distributed Diagnosis and Healthcare (D2H2)*, pages 5–7, Arlington, VA, April 2006.

[84] R. Kapur, T.W. Williams, and E.F. Miller. System Testing and Reliability Techniques for Avoiding Failure. *IEEE Computer*, 29(11):28–30, November 1996.

[85] B. Karaguzel, C.R. Merritt, T.H. Kang, J. Wilson, P. Franzon, H.T. Nagle, E. Grant, and B. Pourdeyhimi. Using Conductive Inks and Non-Woven Textiles for Wearable Computing. In *Proceedings of the 2005 Textile Institute Worlsd Conference*, Raleigh, NC, March 2005.

[86] R. Karlsson, T. Schön, and F. Gustafsson. Complexity Analysis of the Marginalized Particle Filter. *IEEE Transactions on Signal Processing*, 53(11):4408–4411, 2005.

[87] G. Karniadakis and R.M. Kirby. *Parallel Scientific Computing in C++ and MPI*. Cambridge University Press, Cambridge, UK, 2002.

[88] G.E. Karniadakis and S. Sherwin. *Spectral/hp Element Methods for Computational Fluid Dynamics*. Oxford University Press, Oxford, UK, 2005.

[89] R.M. Keller. Denotational Models for Parallel Programs with Indeterminate Operators. In E.J. Neuhold, editor, *Formal Descriptions of Programming Concepts*, pages 337–366, Amsterdam, The Netherlands, 1978. North Holland Publishing Co.

[90] K.H. Kim and C. Subbaraman. Fault-tolerant Real-time Objects. *Communications of the ACM*, 40(1):75–82, January 1997.

[91] R. Kimmel and J.A. Sethian. Fast Marching Methods for Robotic Navigation with Constraints. Technical Report Center for Pure and Applied Mathematics, University of California, Berkeley, Department of Mathematics, May 1996.

[92] V. Klumpp, F. Sawo, U.D. Hanebeck, and D. Fränken. The Sliced Gaussian Mixture Filter for Efficient Nonlinear Estimation. In *11th International Conference on Information Fusion (Fusion 2008)*, Cologne, Germany, 2008.

[93] J. Koenderink. The Structure of Images. *Biol. Cyber.*, 50:363–370, 1984.

[94] D. Komatitsch, J.-P. Vilotte, R. Vai, J.M. Castillo-Covarrubias, and F.J. Sanchez-Sesma. The Spectral Element Method for Elastic Wave Equations - Application to 2-D and 3-D Seismic Problems. In *International Journal for Numerical Methods in Engineering*, volume 45, pages 1139–1164, 1999.

[95] J. Kondo and T. Yamazaki. A Prediction Model for Snowmelt, Snow Surface Temperature and Freezing Depth Using a Heat Balance. *Journal of Applied Meteorology*, 29:375–384, 1990.

[96] P. Krishna, N.H. Vaidya, M. Chatterjee, and M. Steenstrup. A Cluster-based Approach for Routing in Dynamic Networks. *ACM SIGCOMM Computer Communication Review*, 27(2):49–65, April 1997.

[97] T.C. Lacalli and L.G. Harrison. Turing's Conditions and the Analysis of Morphogenetic Models. *Jnl of Theoretical Biology*, 76:419–436, 1979.

[98] A.M. Law and W.D. Kelton. *Simulation Modeling and Analysis*. McGraw-Hill, New York, NY, 2000.

[99] M. Lehning, P. Bartelt, R. Brown, and C. Fierz. A Physical SNOWPACK Model for the Swiss Avalanche Warning Part III: Meterological Forcing, Thin Layer Formation and Evaluation. *Cold Regions Science and Technology*, 35:169–184, 2002.

[100] M. Lehning, P. Bartelt, R. Brown, C. Fierz, and P. Satyawali. A Physical SNOWPACK Model for the Swiss Avalanche Warning Part II: Snow Microstructure. *Cold Regions Science and Technology*, 35:147–167, 2002.

[101] A. Lim. Support for Reliability in Self-Organizing Sensor Network. In *Proc of the Intnl Conf on Information Fusion*, Annapolis, Maryland, July 2002.

[102] J.T.-H. Lo and Jr. S.L. Marple. Observability Conditions for Multiple Signal Direction Finding and Array Sensor Localization. *IEEE-T Signal Processing*, 40(11):2641–2650, 1992.

[103] J. Luitjens, M. Berzins, and T.C. Henderson. Parallel Space-Filling Curve Generation Through Sorting. *Concurrency and Computation: Practice and Experience*, 19(10):1387–1402, 2007.

[104] N. Lynch. *Distributed Algorithms*. Morgan-Kaufman Pub, San Francisco, CA, 1996.

[105] S. Madden, M.J. Franklin, and J.M. Hellerstein. TAG: a Tiny Aggregation Service for Ad-Hoc Sensor Networks. In *Proc. of the Fifth Symposium on Operating Systems Design and Implementation*, Boston, MA, December 2002.

[106] P.K. Maini and H.G. Othmer. *Mathematical Models for Biological Pattern Formation*. Springer-Verlag, Berlin, 2001.

[107] A. Mainwaring, J. Polastre, R. Szewczyk, D. Culler, and J. Anderson. Wireless Sensor Netwroks for Habitat Monitoring. In *WSNA 2002*, Atlanta, GA, September 2002.

[108] J.E. Marsden and A.J. Tromba. *Vector Calculus*. W.H. Freeman and Company, New York, NY, 1988.

[109] H. Meinhardt. *Models of Biological Pattern Formation*. Academic Press, London, UK, 1982.

[110] C.R. Merritt, B. Karaguzel, T.H. Kang, J. Wilson, P. Franzon, H.T. Nagle, B. Pourdeyhimi, and E. Grant. Electrical Characterization of Transmission Lines on Specific Non-Woven Textile Substrates. In *Proceedings of the 2005 Textile Institute Worlsd Conference*, Raleigh, NC, March 2005.

[111] J. Murray. *Mathematical Biology*. Springer-Verlag, Berlin, 1993.

[112] D. Nadig, S.S. Iyengar, and D.N. Jayasimha. New Architecture for Distributed Sensor Integration. In *IEEE SOUTHEASTCON Proceedings*, Charlotte, NC, 1993.

[113] R. Nagpal. Programmable Pattern-Formation and Scale-Independence. In *Proc International Confernce on Complex Systems (ICCS)*, Nashua, NH, June 2002.

[114] B.C. Ng and A. Nehorai. Active Array Sensor Localization. *IEEE-T Signal Processing*, 44:309–327, 1995.

[115] S.-Y. Ni, Y.-C. Tseng, Y.-S. Chen, and J.-P. Sheu. The Broadcast Storm Problem in a Mobile Ad Hoc Network. In *Proceedings of Mobicom 1999*, Seattle, WA, August 1999.

[116] G. Nicolis and I. Prigogine. *Exploring Complexity: an Introduction*. W.H. Freeman and Co, New York, NY, 1989.

[117] K.J. Nurmela and P.R.J. Ostergard. Covering a Square with up to 30 Equal Circles. Technical Report HUT-TCS-A62, Helsinki University of Technology, Laboratory for Theoretical Computer Science, Helsinki, Finnland, 2000.

[118] W.L. Oberkampf, T.G. Trucano, and C. Hirsch. Verification, Validation and Predictive Capability in Computational Engineering and Physics. In *Foundations for Verification and Validation in the 21st Century Workshop*, Laurel, Maryland, October 2002.

[119] P. Perona, T. Shiota, and J. Malik. Anisotropic Diffusion. In B. Romeny, editor, *Geometry-Driven Diffusion in Computer Vision*, Dordrecht, The Netherlands, 1994. Kluwer.

[120] A. Perrig, R. Szewczyk, V. Wen, D. Culler, and J.D. Tygar. SPINS: Security Protocols for Sensor Networks. *Wireless Networks*, 8(5):521–534, Sept 2002.

[121] L. Prasad, S.S. Iyengar, R.L. Kashyap, and R.N. Madan. Functional Characterization of Fault Tolerant Integration in Distributed Sensor Networks. *IEEE Transactions on Systems, Man and Cybernetics*, 21(5):1082–1087, September 1991.

[122] L. Prasad, S.S. Iyengar, R.L. Rao, and R.L. Kashyap. Fault-tolerant Sensor Integration using Multiresolution Decomposition. *Journal of The American Physical Society*, 49(4):3452–3461, April 1994.

[123] I. Prigogine. *Thermodynamics of Irreversible Processes*. Interscience Publishers, New York, NY, 1968.

[124] I. Prigogine. *From Being to Becoming: Time and Complexity in the Physical Sciences*. W.H. Freeman and Co, San Francisco, CA, 1980.

[125] J.A. Profeta. Safety-critical Systems Built with COTS. *IEEE Computer*, 29(11):54–60, November 1996.

[126] B. Randell. System Structure for Software Fault Tolerance. In R.T. Yeh, editor, *Current Trends in Programming Methodology, Vol. 1*, pages 195–219, Englewood Cliffs, NJ, 1977.

[127] K. Rankinen, T. Karvonen, and D. Butterfield. A Simple Model for Predicting Soil Temperature in Snow-covered and Seasonally Frozen Soil: Model Description and Testing. *Hydrology and Earth System Sciences*, 8:706–716, 2004.

[128] K. Roberts and U.D. Hanebeck. Prediction and Reconstruction of Distributed Dynamic Phenomena Characterized by Linear Partial Differential Equations. In *Proceedings of the 8th International Conference on Information Fusion (Fusion 2005)*, Philadelphia, Pennsylvania, July 2005.

[129] S.M. Ross. *Simulation*. Elsevier, Amsterdam, The Netherlands, 2006.

[130] L. A. Rossi, B. Krishnamachari, and C.-C.J. Kuo. Distributed Parameter Estimation for Monitoring Diffusion Phenomena Using Physical Models. In *First Annual IEEE Communications Society Conference on Sensor and Ad Hoc Communications and Networks (SECON 2004)*, pages 460–469, Los Angeles, USA, 2004.

[131] F. Sawo, F. Beutler, and U.D. Hanebeck. Decentralized Reconstruction of Physical Phenomena based on Covariance Bounds. In *Proceedings of the 17th IFAC World Congress (IFAC 2008)*, Seoul, Republic of Korea, July 2008.

[132] F. Sawo, T.C. Henderson, C. Sikorski, and U.D. Hanebeck. Sensor Node Localization Methods based on Local Observations of Distributed Natural Phenomena. In Sukhan Lee, editor, *Proceedings of the IEEE International Conference on Multisensor Fusion and Integration*, Seoul, South Korea, August 2008. IEEE.

[133] F. Sawo, M.F. Huber, and U.D. Hanebeck. Parameter Identification and Reconstruction Based on Hybrid Density Filter for Distributed Phenomena. In *10th International Conference on Information Fusion (Fusion 2007)*, Quebec, Canada, July 2007.

[134] F. Sawo, V. Klumpp, and U.D. Hanebeck. Simultaneous State and Parameter Estimation of Finite-Dimensional Models of Distributed Systems based on Sliced Gaussian Mixture Filter. In *11th International Conference on Information Fusion (Fusion 2008)*, Cologne, Germany, 2008.

[135] F. Sawo, K. Roberts, and U.D. Hanebeck. Bayesian Estimation of Distributed Phenomena Using Discretized Representations of Partial Differential Equations. In *Proceedings Proceedings of the 3rd International Conference on Informatics in Control, Automation and Robotics (ICINCO 2006)*, pages 16–23, Setubal, Portugal, August 2006.

[136] F. Sawo, K. Roberts, and U.D. Hanebeck. Bayesian Estimation of Distributed Phenomena using Discretized Representations of Partial Differential Equations. In *3rd International Conference on Informatics in Control, Automation and Robotics (ICINCO 2006)*, pages 16–23, Setubal, Portugal, August 2006.

[137] L. Schenkat, L. Veigel, and T.C. Henderson. EGOR: Design, Development, Implementation – An Entry in the 1994 AAAI Robot Competition. Technical Report UUCS-94-034, University of Utah, Salt Lake City, UT, December 1994.

[138] V. Schnayder, M. Hempstead, B. Chen, G.W. Allen, and M. Welsh. Simulating the Power Consumption of Sensor Network Applications. In *ACM SenSys Proceedings*, Baltimore, MA, November 2004.

[139] S.A. Schneider, V. Chen, and G. Pardo. ControlShell: A Real-time Software Framework. In *AIAA Conference on Intelligent Robots in Field, Factory, Service and Space*, Dayton, OH, 1994.

[140] T. Schön, F. Gustafsson, and P.-J. Nordlund. Marginalized Particle Filters for Nonlinear State-space Models. Technical report, Linköpings–University, 2003.

[141] O.C. Schrempf and U.D. Hanebeck. A State Estimator for Nonlinear Stochastic Systems Based on Dirac Mixture Approximations. In *Proceedings of the 4th International Conference on Informatics in Control, Automation and Robotics (ICINCO 2007)*, volume SPSMC, pages 54–61, Angers, France, May 2007.

[142] J.A. Sethian. *Level Set Methods and Fast Marching Methods*. Cambridge University Press, New York, 1996.

[143] E. Shilcrat, P. Panangaden, and T.C. Henderson. Implementing Multi-sensor Systems in a Functional Language. Technical Report UUCS-84-001, The University of Utah, Salt Lake City, UT, February 1984.

[144] K. Shin, A. Abraham, and S.Y. Han. Self-Organizing Sensor Networks using Intelligent Clustering. In *LNCS Proceedings of the Workshp on Ubiquitous Web systems and Intelligence*, Berlin, Germany, 2006.

[145] D. Simon, B. Espiau, E. Castillo, and K. Kapellos. Computer-Aided Design of a Generic Robot Controller Handling Reactivity and Real-time Issues. *IEEE Transactions on Control Systems Technology*, 4(1), 1993.

[146] R. Smith, A. Frost, and P. Probert. A Sensor System for the Navigation of an Underwater Vehicle. *The International Journal of Robotics Research*, 18(7):697–710, July 1999.

[147] A.J. Sommese and II C. Wampler. *The Numerical Solution of Systems of Polynomials Arising in Engineering and Science*. World Scientific, New York, NY, 2005.

[148] H.W. Sorenson. *Kalman Filtering: Theory and Application*. Piscataway, NJ: IEEE, 1985.

[149] K. Stephenson. *Introduction to Circle Packing*. Cambridge University Press, New York, NY, 2005.

[150] D.B. Stewart and P.K. Khosla. Mechanisms for Detecting and Handling Timing Errors. *Communications of the ACM*, 40(1):87–93, January 1997.

[151] M. Stoffel. *Numerical Modelling of Snow Using Finite Elements*. PhD thesis, Swiss Federal Institute of Technology Zürich, Zürich, Switzerland, May 2005.

[152] V. Swaminathan, K. Chakrabarty, and S.S. Iyengar. Dynamic I/O Power Management for Hard Real-time Systems. In *Proc. Intl. Symposium on Hardware/Software Co-Design (CODES*, pages 237–242, Ambleside, Lake District, UK, 2001.

[153] P.G. Szabo, M.C. Markot, T. Csendes, E. Specht, L.G. Casado, and I. Garcia. *New Approaches to Circle Packing in a Square*. Springer, New York, NY, 2007.

[154] D.G. Thaler and C.V. Ravishankar. Distributed Top-Down Hierarchy Construction. In *Proceedings of INFOCOM 1998*, pages 693–701, San Francisco, CA, 1998.

[155] P.F. Tsuchiya. The Landmark Hierarchy: a New Hierarchy for Routing in Very Large Networks. *ACM SIGCOM Computer Communication Review*, 18(4):35–42, August 1988.

[156] A. Turing. The Chemical Basis of Morphogenesis. *Philosophical Transactions of the Royal Society of London*, B237:37–72, 1952.

[157] D. Ucinski. *Measurement Optimization for Parameter Estimation in Distributed Systems*. Technical University Press, Zielona Gora, Poland, 1999.

[158] H. Wang, H. Lenz, A. Szabo, and U.D. Hanebeck. Fusion of Barometric Sensors, WLAN Signals and Building Information for 3–D Indoor/Campus Localization. In *Proceedings of the 2006 IEEE International Conference on Multisensor Fusion and Integration for Intelligent Systems (MFI 2006)*, pages 426–432, Heidelberg, Germany, September 2006.

[159] G.A. Weller, F.C.A. Groen, and L.O. Hertzberger. A Sensor Processing Model Incorporating Error Detection and Recovery. In *Traditional and Non-traditional Robotic Sensors* Edited by T. C. Henderson, pages 351–363. Springer-Verlag, Berlin, 1990.

[160] K. Whitehouse. The Design of Calamari: an Ad Hoc Localization System for Sensor Networks. Master's thesis, University of California, Berkeley, San Francisco, CA, 2002.

[161] K. Whitehouse and D. Culler. Calibration as Parameter Estimation in Sensor Networks. In *Proc. WSNA 2002*, Atlanta, GA, September 2002.

[162] K. Whitehouse and D. Culler. Macro-Calibration in Sensor/Actuator Networks. In *Proceedings of Mobile Networks and Applications*, volume 8, pages 463–472, 2003.

[163] N. Wirth. On the Compostion of Well-Structured Programs. In E.N. Yourdan, editor, *Classics in Software Engineering*, pages 153–172, London, 1979.

[164] J.D. Wright. Measurements, Transmission, and Signal Processing. In D.A. Mellichamp, editor, *Real-Time computing*, pages 80–112, New York, NY, 1983.

[165] Y. Xu, S. Bien, Y. Mori, J. Heidemann, and D. Estrin. Topology Control Protocols to Conserve Energy in Wireless Ad Hoc Networks. Technical Report 6, University of California, Los Angeles, Center for Embedded Networked Computing, January 2003.

[166] Y. Yemini, S. da Silva, D. Florissi, and H. Huang. The Network Flow Language: A Mark-based Approach to Active Networks. Computer Science XXX, Columbia University, July 1999.

[167] L. Zhang. Simple Protocols, Complex Behavior. In *Proc. IPAM Large-Scale Communication Networks Workshop*, March 2002.

[168] F. Zhao and L. Guibas. *Wireless Sensor Networks*. Elsevier, Amsterdam, 2004.

[169] T. Zhao and A. Nehorai. Detecting and Estimating Biochemical Dispersion of a Moving Source in a Semi-Infinite Medium. *IEEE Transactions on Signal Processing*, 54(6):2213–2225, June 2006.

[170] S.C. Zhu and D. Mumford. Prior Learning and Gibbs Reaction-Diffusion. *IEEE-T on Pattern Analysis and Machine Intelligence*, 19(11):1236–1250, 1997.

[171] Perfect Ice Conditions Ensure Faster Speed Skating Times (url: http://www.zigbee.org/imwp/download.asp?ContentID=12588). *ZigBee Alliance*, 2008.

Index